Policies on Health and Safety in Thirteen Countries of the European Union

Volume II: The European Situation

EF/96/14/EN

Policies on Health and Safety in Thirteen Countries of the European Union

Volume II: The European Situation

Professor Françoise Piotet

 European Foundation
for the Improvement of Living and Working Conditions
Loughlinstown, Dublin 18, Ireland
Tel: +353 1 204 3100 Fax: +353 1 282 6456

Cataloguing data can be found at the end of this publication

Luxembourg: Office for Official Publications of the European Communities, 1996

ISBN 92-827-6642-X

Printed in Ireland

Preface

The European Work Environment Survey published by the Foundation in 1992 showed important differences among member states in working conditions related to health and safety. These inequalities persist even when economic factors are taken into account. Possible explanations of these imbalances include the health and safety strategies developed by public authorities and companies in the member states.

In 1993 the Foundation launched a project on identification and assessment of occupational health and safety strategies in Europe to produce an overview of the situation in the member states. National reports were contracted in 13 member states and two European reports were prepared. The first one describes the health and safety policies of each Member State and is based on the 13 leaflets that summarise the national reports. The second one, that you have in your hands, develops a more analytic approach. The two are complementary and will be presented as volumes I and II of the same study.

This volume has been written by Prof. Françoise Piotet and is based on 13 national reports. The complexity of the health and safety systems is high and they are in almost continuous change. Therefore inaccuracies can easily slip in the text without being noticed except by national experts. So the report has been reviewed and commented by the authors of the national reports and evaluated by a committee of representatives of national governments, European employers and trade unions representatives and the European Commission.

We are grateful to the authors of the 13 national reports, the members of the evaluation committee and especially to the author.

We hope this volume will bring new ideas and discussions towards an Europe with better health and safety.

Clive Purkiss
Director

Eric Verborgh
Deputy Director

TABLE OF CONTENTS

VIII

LIST OF NATIONAL REPORTS

Belgium	**V. de Broeck, M. de Greef, M. Heselmans, M. van der Steen**, ANPAT, Brussels
Denmark	**P.B. Olson**, Department of Environment, Technology and Social Studies, Roskilde University
Finland	**J. Rantanen**
France	**Prof. F. Piotet**, Laboratoire Georges Friedmann, Paris
Germany	**B. Badura**, University of Bielefeld **F. Schlottman**, School of Public Health **C. Kuhlmann**, Department of Social Epidemiology and Health Systems Design
Greece	**I. Banoutsos, S. Papadopoulos, N. Sarafopoulos, E. Velonakis, C. Marouli**, Ergonomia Ltd.
Ireland	**E. McCarthy, M.T. Byrne**, Social and Organisational Psychology Research Centre, University College, Dublin
Italy	**G. Frigeri**, SNOP
Netherlands	**P. Smulders, M. van Domolen, E. de Gier, M. Kompier, R. de Winter**, TNO Prevention and Health, Division Work and Health, Leiden
Portugal	**S.M. Cardoso**, Instituto de Higiene e Medicina Social, Faculdade de Medicina de Coimbra
Spain	**Prof. L. Lemkow**, Universitat Autonoma de Barcelona
Sweden	**I. Stymne**
United Kingdom	**D. Walters**

INTRODUCTION

The European Foundation for the Improvement of Living and Working Conditions launched, at the end of 1993, in the 13 European member states or future member states at that time of the EU, a survey to describe, based on as factual an account as possible, the general situation in relation to health and safety in each member state. This description would cover all their legal, regulatory and contractual provisions, together with their implementation of the European Community directives on this subject. It would thus present the various actors who play a part, in different ways, within the particular structures of each country, the structures responsible for implementing policies at all levels, the supervisory and promotional structures, and the systems of insurance and compensation. The means of research and the priorities in this field may well contribute to clarifying the problems confronting the various countries and the means employed for finding solutions to them.

The statistics which have been supplied in respect to industrial accidents and occupational diseases give an idea of the results of the efforts made at all levels in matters of health and safety. It would, nevertheless, be quite wrong to take these figures as the only pertinent indicator of the effectiveness of the measures employed and many reasons militate in favour of qualifying any such judgment. The methods of compiling such figures vary greatly from country to country and affect the accuracy of the results. A rise in certain figures here or there may, paradoxically, signify an improvement in the supervisory system rather than a disimprovement in the situation. In every country, production systems are in a process of transformation which contributes to a modification of the structure of the working population and a shifting of the risks. Such changes naturally affect the results in this domain and for this reason their context has been briefly described in each of the reports.

Each rapporteur was asked, in addition to data which is as factual as possible, to give the opinions of those who are particularly competent in the field of health and safety: members of employer organisations and trade unions, and representatives of the public service and the universities. These views of a more perspective nature can only provide enlightenment to allow us to discern trains of thought about the future.

• The research plan

In order to obtain a satisfactory result from this study, the European Foundation established a plan of research which allowed it to be carried out in stages.

1)   An initial series of working sessions between the author of this report and the Foundation project managers in charge of this portfolio, Jaume Costa and Pascal Paoli, helped both to pinpoint the objective sought and to draw up a preliminary schedule to serve as the framework for the collection of data. The thirteen countries contracted to participate in this study are: Belgium, Denmark, Finland, France, Germany, Greece, Ireland, Italy, the Netherlands, Portugal, Spain, Sweden and the United Kingdom.

2)   During one long working session, this schedule was presented to all the rapporteurs to give them an opportunity to comment upon or amend it if necessary (September 1993).

3)   Having incorporated the rapporteurs' comments, a final framework was established and a relatively short time fixed for the completion of the reports. A working meeting in September 1994 allowed each one to present the main results and to make suggestions in regard to the consolidated report (September 1994).

Naturally these reports were sent to the various national authorities concerned, some of whom did not fail to react to the texts proposed by the researchers. The final versions of some reports were therefore not sent in until the very end of 1994. The framework supplied, which was used to structure the reports, was not intended to formalise the final reports too strictly. Relative to the data available, time spent on writing summaries and the inevitable feelings which this very difficult subject arouses, the national reports were finally quite diverse but supplied, on the whole, a solid basis of reference.

The time for summarising is the time for discovering any discrepancies which can only be attributable to the authors of the framework and in particular to the author of this report.

• The philosophy of the consolidated report

A summary is not a comparison. It is not the purpose of this report in any way to make a comparison of what is not comparable. The differences in policies and structures, and in the place accorded to health and safety in the workplace, depend upon a multiplicity of complex factors which are often the products of a long and varied history. The thirteen countries which participated in this study represent a working population in the vicinity of 146 million workers and cover an area which extends from the Arctic to the borders of the Mediterranean. If we agree to characterise developments by criteria which have to do with types of industrialisation, it is obvious that there are still very great differences between, for example, Greece and Sweden. Comparing accident statistics without taking into account the effects of structure and scale would, strictly speaking, make no sense (although this is often done, since figures, in this field, play a decisive role).

Without venturing into the hazards of an analysis in terms of national culture and keeping to the heart of our subject, it is obvious that history and geography have been responsible for giving each of the countries concerned an individual face. The right to work, in particular, has emanated in these countries from very different legal sources (Romanist or Germanic) which have had a decisive influence not only on the way in which health and safety problems are approached but also on the concept of work, on enterprise and on the place and status accorded to human beings in their work.

Cultures of confrontation and distrust cohabit with the old traditions of cooperation and this undoubtedly has its impact on working conditions.

In order to avoid the risk of a comparison but also of a probably rather fastidious enumeration, it has been decided, in drawing up this consolidated report, not to follow strictly the plan selected for the national summaries. Having taken the main elements from the national reports which allow us to draw the broad outlines (in Chapter 1) of the socio-economic context surrounding the health and safety policies but also allows us to explain to some extent the constraints and difficulties met with, we then concentrated upon the structures (Chap. 2), the actors (Chap. 3) and the policies (Chap. 4). Chapter 5 is devoted to a summary of the positions of the experts consulted by the rapporteurs in each country. **Within this classification, we have attempted to distinguish systematically between the structures and the company's internal and external actors.**

This presentation does not claim to be exhaustive, much as we should like it to. Only the data supplied in the national reports have been used for this summary. In the time allotted, it was scarcely possible to fill the information gaps.

CHAPTER 1

THE CONTEXT

The contexts in which the policies on health and safety have been analyzed are very varied. Certain points which one hopes are pertinent have been selected to characterise them. Firstly the structure of the population will be described in a context of growing unemployment. Locations and types of employment have undergone changes in recent times which must be taken into account in order to appreciate the evolution in industrial hazards. The system of industrial relations itself is directly affected by these upheavals. In some countries, its weakening poses special problems for the implementation of policies to improve working conditions.

1.1 The structure of the working population

The figures presented by the various countries regarding their working population and its structure, together with the rate of unemployment, have been assembled in the following table.

Table 1: Working populations distributed by sector and rate of unemployment

Country[1]	Working populati- on[2]	Agri- cul- ture[3]	Indus- try[3]	Ser- vices[3]	Unemploym- ent[3]
Germany (90)	37.445	3.6	40.5	55.8	8,(14)[4]
Belgium (91)	4.210	0.7	30.8	68.5	11.8
Denmark (93)	2.507	5	26.6	68.4	12.4
Spain	13.100	7.8	38.4	53.8	17.3
Finland	2.500	9	27	64	20
France (90)	25.103	6.1	29.2	64.6	12.2 (93)
U.K. (94)	20.888	1.2	25.5	73.3	9.9
Greece (91)	3.632	22.2	27.4	50.2	9.7
Italy (93)	20.000	7.6	33	59.4	6
Ireland (92)	1.139	13.4	28	58.6	15.8
Netherlands (93)	5.925	4	25	71	5
Portugal (93)	4.542	11.3	32.9	55.8	5.4
Sweden (94)	3.926	3.4	25	71.6	8

In its summarised form, this table gives an initial idea of the diversity of the contexts which characterise the countries in which the investigation into policies on health and safety at work was carried out.

The structure of the working population demonstrates the relatively different types of development in each country, which are not all linked to the length of time taken by such development. They show, in every case, the special character- istics of their productive system. In the case of countries

---

[1]   The figure in brackets represents the year of reference for the statistics on the working population

[2]   In millions

[3]   In percentages

[4]   The two figures represent unemployment figures in the former E. Germany and those for the new Länder

the volume of whose active working population is the largest
and which, in the main, have a long industrial history, the
differences are even more significant. Compared, for example,
with the British or French situations, the secondary sector in
Germany still occupies an exceptional place. The volume of
agriculture is very variable from country to country.

Some of the reports supplied information not only on the
present state of this structure but also on its evolution.
Certain countries appear to have achieved a relative balance
and the population transfers between sectors are now slower.
Others, on the contrary, are still evolving as is the case,
for example, in Portugal where the agricultural workforce is
rapidly diminishing.

This initial division by sector allows us to form hypotheses,
not on the situation regarding health and safety, but on the
nature of the problems one is in danger of encountering. It
would be necessary to have more specific data in order to
characterise each of these sectors. Thus the mechanisation of
agriculture and the use of chemicals appear as factors
entailing special risks in regard to accidents at work or
occupational diseases. In the secondary sector, the transformations are also very rapid. The German report emphasises,
for example, that in manufacturing employment, almost 35% of
the jobs are tertiary sector jobs. When they were supplied,
the statistics concerning the various occupational categories
showed, in the majority of cases, a significant reduction in
manual jobs and an increase in those in engineering and
management, but also of workers.

On the other hand - and the term used here takes on new
meaning in relation to its accepted one - the tertiary is
being partially industrialised. Machines of a new kind but
also the more traditional ones are penetrating this sector and
new hazards are appearing here. Some examples of this are in

the health sector and also in the wholesale distribution sector.

While keeping to the categories used by the economists to characterise the productive system, a strong tendency towards the development of the service sector at the expense of two other sectors is, however, evident. To this should be added (and reference should be made to the national reports for the quantitative figures which have not been included, since the definitions of the public sector are too disparate from one country to another) the increase in public sector jobs that are dependent on the State or on local authorities.

These most important distinctions, which are too readily evoked, probably show that the categories used to describe the context may be insufficient to allow identification of the difficulties which may be met in the fields of health and safety.

Here, a second observation should be made. In the great majority of the countries which were the subject of this research the principal measures taken in the matter of health and safety, even if some of these originated at the turn of the century, were fully developed in the period of the "Glorious Thirties"[5], in other words, at the height of industrialisation. The question which is now pertinent is the adaptation of the existing health and safety measures to a reality which has profoundly changed.

1.2 Unemployment and the changes in the working population

1.2.1 Unemployment
As the last column of Table 1 indicates, the rate of unemployment, with few exceptions and assuming the accuracy of the figures available to us, is relatively high and what chiefly

---

[5]   Name given by the economists (J. Fourastié) to the period of
       thirty years' growth which followed the end of World War II.

characterises it is its growth in the course of the previous decade. In regard to the subject of our interest, it is easy to imagine that unemployment can have a multiplicity of consequences. In many countries, unemployment, apart from its volume, mainly affects the populations situated at the two extremes of the age pyramid, but also women. It is then - and here again the variations between countries are important - very often a case of long-term unemployment. Finally, in countries of high immigration, it may affect these populations in particular.

The high rate of unemployment naturally leads the authorities to give priority to getting people back to work. The risk is that this priority, whose legitimacy is incontestable, may relegate working conditions and their effects on the health of workers into second place. Besides, the effects of unemployment on the budget and the costs of its social management may lead the authorities to reduce expenditure on areas where the immediate gains are the least visible, research and also education being the first to come under threat.

One may also observe, as certain reports have emphasised, (United Kingdom, France, Spain and Sweden) a relaxation of vigilance by the social partners, but also by the controlling body on the question which concerns us in favour of this priority given to employment.

The existence of such high unemployment naturally impacts primarily on the unemployed themselves. Apart from its economic dimension, information is beginning to be available which shows the impact of long-term unemployment on health. One may also, as some national reports do, point to the impact on their attitude to work, acceptance of poor working conditions and the stress under which the threat of unemployment places a growing number of workers who feel their jobs are increasingly at risk.

1.2.2 Changes in the working population

In this section we shall remind ourselves of the changes which have taken place in the population as a whole and whose repercussions have been felt on the working population, its structure and its collective behaviour.

The first significant characteristic to be taken from the national studies is the evolution of the **demographic structure**, indicated in all the countries, including the most southerly, by an ageing of the working population. Apart from the consequences which this might have on retirement schemes, this ageing, which can only increase in the coming decades taking into account the present birth rate and the fact that young people are entering the workforce later and later, is a major element to be considered in policies to be pursued on the subject of health and safety (cf. the insistence on this point made by the report on the Netherlands). One can see, in this regard, the major role to be played by prevention policies and the essential part played by working conditions and the organisation of labour.

The second characteristic emphasised by the national reports is the contribution made by **women** in the workplace. In the case of the date available to us (eight reports) the lowest figure here is for Ireland, with a feminisation rate of 32.8% of the working population. In the case of countries with a long industrial history this figure oscillates around 40%. The countries whose working populations are least feminised are also showing a trend towards the growth of feminist activity. Few of the reports supply detailed figures on female employment but its importance will doubtless necessitate special reflection upon the consequences of the working conditions to which women are subjected. In addition, all the studies emphasise the segregation which continues to operate between men and women in the workplace.

Information of any precise nature in relation to the **immigrant population** figures in only a small number of reports. Some of the countries on the list, like Ireland, Spain and Portugal, have been areas of emigration, whereas the United Kingdom, France, Belgium, Germany and the Netherlands, both for historic reasons but also because of the nature of their industrialisation, have acted as host to immigrant workers from relatively varied geographic locations. From the information available to us it appears that in countries with high immigration, the proportion of foreign workers in the working population varies between 5% and 7%. Failing precise information on occupational diseases for this category of workers, it is obvious that the high proportion of industrial accidents which concern them refers mainly to the sectors which employ them in the greatest numbers, in particular, the building trade, public works and the docks.

1.3 Locations and types of employment

All the national reports evidence one characteristic too often forgotten: even if they tend to occupy an important part of the working population, **large companies are an exception.** In European countries, the norm is chiefly the existence of a multiplicity of small or very small businesses. It has not been possible, in the form of a table, to summarise the data referring to company size since the scales of measurement used in presenting these figures are not the same for every country. Failing easily accessible detailed data in each of the national reports, it would appear that more than half, and in many cases, more than three-quarters of the working population are employed by companies with less than 100 workers. Now - and we shall see this in the section of this report devoted to structures - the collective measures in respect of health and safety at company level have, as a general rule, been designed for medium-sized or large companies. By the same token, union representation has - with exceptions - been mainly developed in large companies. In the years to come, a study of the means to implement the development of active policies on health and safety in small companies should be a major preoccupation.

The transformations in the economic environment of European businesses, the important technological developments which have led some to refer to a third industrial revolution, but also the crisis which affects to different degrees all of our countries, contribute to changing - whether temporarily or definitively - our traditional forms of employment.

Part-time work is increasing. The rate is higher in the northern countries than in southern Europe but, in the latter, trends show a rise which is tending to equal the former. The highest percentage here is that reached by the Netherlands (29%) and the lowest by Portugal (6.5%). (Let us remind ourselves here that this type of employment has only been authorised in Greece since 1990). In addition, the definition of part-time work varies from country to country and, in particular, there is a great diversification of this part time which can no longer be assimilated into a half-day job. Because it is essentially female employment, because, as in the United Kingdom, it often leads to double jobbing (according to the rapporteur, 12,25% of the working population have a second job), and because these part-time jobs do not offer the same career opportunities as permanent jobs, this type of employment has often been the subject of controversy. From the point of view of health and safety, it deserves much closer observation than it has received up to now.

Self-employment is, according to the two rapporteurs, very common in Greece and Portugal (21.8% and 24.7% respectively). It seems, although all the data is not available to confirm this, that this type of employment has tended, with the crisis, to grow as well in countries with a long industrial tradition. The means to supervise the working conditions and health of the self-employed are, of course, most problematical.

All the reports, without always supplying all the quantitative data which would allow a precise evaluation of the scale of the phenomenon, castigate the development of **job insecurity**. Here again, comparisons are very difficult due to the absence of a uniformity of definitions and the great diversity of forms this takes. Depending on the form it takes, it does not, of course, have the same effect on the health and safety of workers. **Temporary work**, as is now the case in France but also in other EU countries, is subject to rules for specific controls as is often the case, too, in relation to **specified term contracts**. But what is the reality in following up the health of workers when mobility is so high? Increasing **sub-contracting** is another form of flexibility to which large companies have recourse, at times imposing very difficult working conditions on their sub-contractors. All the **work experience contracts** or back-to-work schemes for the long-term unemployed have also contributed to a confusion of the exact categories on the basis which we used to analyze jobs and stipulate the health and safety regulations which should apply to them. This destabilisation of the traditional forms of employment has not reached the same proportions in each of the countries, nor is it likely to have had the same effects on the populations concerned. However, the very great difficulties which young people from many European countries have in obtaining long-term employment and the multiplication of insecure jobs, threaten to weigh heavily upon the future.

Finally, this very brief sketch must include - even though they are treated incidentally in the reports - the changes in work in the three sectors concerned. Agriculture has been mechanised. The 'machine' and products used mean that agricultural workers incur the same risks which used to be typical of industry. Large companies are changing just like the services sector due to organisational changes but also to the introduction of new technologies which can profoundly alter man's relationship with his work. The body takes fewer risks but fatigue and stress take other forms which, because they are less visible, are more difficult to perceive and take into account. In the construction and public works sector - in spite of the efforts made in regard to inspection and the introduction of new regulations - the classic risks remain. If in certain cases working conditions are improving, in others it is obvious they are deteriorating. **Shift work**, which saw a certain decline at the end of the Eighties, partly due, doubtless, to the vigorous campaign with which the Foundation was associated and whose damaging effects were denounced all over Europe, is again on the increase. It is affecting new sectors and itself takes new forms. If working hours are reduced, atypical timetables such as weekend work or shifting times, are increasing. **'Moonlighting'** which, by its nature, is difficult to evaluate, occurs on various scales depending on the country although it is not possible to say with certainty whether the present difficult period is conducive to its growth. The national reports which suggest estimations provide figures whose differences are important. In Belgium, the product of illegal work is estimated at 15% of GNP and represents between 300,000 and 400,000 workers. The same percentage is supplied by the statistics office of the Netherlands, while in Germany, the product of illegal work only represents 1% of GNP. Apart from its volume its existence naturally poses very specific problems of control in matters of health and safety.

1.4 The evolution of the social partners

More than in other spheres, the state of industrial relations in traditionally industrial countries is the fruit of a long history whose knowledge is necessary to its understanding. In all the countries, the economic climate affects these relations but all the actors in the system do not react in the same way.

In regard to the workers' **unions**, the chief discussions are rather pessimistic. Statements refer to the drop in membership numbers and the crisis in their representativeness. The information supplied by the national reports does not confirm this overall assessment which is only partly verified. What the figures show is the very great difference in the unionisation of the countries selected for the survey. Three groups stand out. The first, made up of Belgium, Denmark, Finland and Sweden, report a rate of unionisation oscillating between 80% and 90%, indicating that practically the whole working population is unionised. The second group is represented by Germany and Ireland, with a rate of between 40% and 50%. Greece, the United Kingdom, the Netherlands and Italy report a rate of unionisation of from 20% to 30%. Spain would probably also have to be placed in this group. The national report indicates a rate of 8% which does not correspond to any of the known figures for that country. France, for its part, has the lowest rate of unionisation in Europe. It is currently under 10%. It is interesting to emphasise the contrasting effects of the crisis on membership; in highly unionised countries it seems to have produced a stimulating effect while, on the other hand, accelerating the process of de-unionisation in countries where this was already low.

These figures are certainly an important indication of the strength of the workers' unions but this is not the only one to be taken into consideration in assessing their influential capacity. The union organisations are very different from country to country. The dominant fashion is an organisation by large occupational categories, grouped if necessary within a single confederation. The British and Irish unions are traditionally divided into trades but this has begun to die out as witnessed by the numerous amalgamations of unions. Ideological cleavages, whether religious or political, are in some countries also translated by competitive structures.

In the majority of countries, the rate of membership of **employer organisations** is high, sometimes higher than union membership as is the case in Germany while in France it is very close to union membership. The employer structures are always organised in accordance with company size and occupational sector.

The information furnished by the reports on **collective bargaining** are not sufficient to enable a meaningful summary to be made. In particular, the information on health and safety which is available does not allow us to make any distinction between what the law provides and what is the result of negotiation. Several points may, however, be emphasised. Even while industrial relations rest upon a conflictual tradition, health and safety constitute a strong point of cooperation between the social partners and this is true at all levels. Collaboration within the bodies responsible for managing the insurance schemes (cf. Chap. 2) is good but this is also the case in respect of all the organisms specialising in co-management (the Swedish system is going through a crisis which is questioning the quality of these relations. Will it last?) The social partners, each in so far as it concerns them, are making a great effort to educate their members on these matters.

Even if some elements bring to light common trends, the contextual disparities between the different countries are still very marked and should be kept in mind in order to understand the differences in structures as well as the behaviour of the actors and the choice of policies we shall deal with in the following chapters.

CHAPTER 2

THE STRUCTURES

This chapter will examine the principal structures which intervene in the field of health and safety.

At national level, they essentially concern the services of labour inspection and the insurance system covering occupational risks. At company level, these will be the prevention services and the structures for staff representation. The research, finally, constitutes the last section of the measures which will be examined.

2.1 National systems of inspection and insurance

In the majority of the countries, the structures responsible for the inspection of companies and those which manage the insurance of workers against industrial risks are, generally speaking, quite separate. This is so for obvious reasons to do with their respective tasks but also with the differences in the status of these bodies. The application of the law naturally comes under the ruling power of the State while the insurance systems may be completely private or arise out of the general Social Security system. However, the separation between the two systems is not as strict as one might imagine, since the tasks of both are changing and evolving. The inspection services are now adding to the monitoring which remains the core of their job, counselling, diagnosis and prevention. Some insurance bodies also actively practise the control which is conceived as one of the means for prevention. The representatives of these two structures may meet on the field or find themselves in competition as may be the case in the dual system in Germany. The oldest systems, however, come together on the priority accorded to the prevention of accidents at work and of occupational diseases.

2.1.1   Labour inspection

The structures and working methods used by the labour inspectorate in European countries have already given rise to
several comparative research projects. This information is
now familiar and it is therefore unnecessary to go into a
detailed description of these structures. We shall only refer
to what concerns health and safety in the work of the inspection services.

Between the thirteen countries which are the subject of this
study, there appear to be quite a few common features in the
existing structures. The ratification of the ILO Convention
81 in regard to the role and powers of the inspectorate and
the status of inspectors has been an important factor in
harmonising the structures and functions associated with them.
But differences also exist which ought to be underlined.
These differences have, of course, as much to do with the more
general characteristics of the political and administrative
structures of the countries in question as with the economic
context and, doubtless, with the diversified concepts of the
role of the inspectorate.

A comparison based on the organisation and concept of the
structures which are responsible for inspecting workplaces
shows that they have, in their own ways, attempted to reply to
the same questions posed by the complexity of the causes which
influence conditions of health and safety. The more knowledge
is gained in this field and the more it appears necessary to
proceed with a global approach and less with merely strict
conformity with the regulations or the norm, it appears this
will be sufficient to attain the levels of health and safety
one might expect in the Member States of the European Union.

The attachment of the inspection services to the ministries
responsible for labour is a constant in all the countries,
except in Italy. There, the inspection services are attached
to the Ministry of Health regarding to principles and guideli-

nes, and to Regional Governments from the operative point of view. There may be some exceptions in agriculture, mining, transport, the maritime and port authorities and, where they exist, the nuclear installations, whose activities are characterised by very specific problems. In a certain number of countries the agents of the public service have special treatment. Failing acceptance to exercise control over its own operations, the State employer leaves this task to internal bodies which represent the officials. As several rapporteurs have emphasised, it is not certain that such bodies can fulfil this role, taking into account the powers and means at their disposal.

As a general rule, there is a clearly identified division of structures between those which derive from the insurance and cover of risks and those which come under the more classic heading of inspection. Whether these two functions are under the same ministry and are assigned to different directorates or whether they come under two different ministries, there are nearly always coordinating bodies at least at national and regional level. As an example one may cite Belgium where the Ministry of Labour has two large directorates, one in charge of industrial safety, organised in 12 departments, and the other concerned with hygiene and industrial medicine, with 5 departments, coordination between the two services being carried out by an 'integrating manager'. In France, coordination between the two ministries responsible for labour and health and social security is arranged at national and regional level by 'coordination committees'. There are some notable exceptions to this principle. Germany is a good example, where they have a dual system.

- **Regional organisation**

At regional level (and the definition of this geographic and administrative entity varies from country to country) we find the expert services which may be called on to support the inspectorate in its work. These services are more or less developed according to country: an engineer (though not in every region) and a doctor in France where, it is true, the inspectorates also have the benefit of specialised services from the National Health Insurance Fund (CNAM) and, if necessary, from the laboratories of the National Institute for Research and Safety (INRS). More highly developed services have been created in the United Kingdom in each of the seven administrative regions, i.e. the Field Consultants Groups and the Field Scientific Support Groups.

The regional level intermediaries for certain functional support services do not exist in all the countries, certainly not in the smallest of them.

• **Local level**

It is at local level, in the performance of the more practical jobs assigned to the inspectors that the greatest differences between the countries are to be found. It is particularly interesting to examine the local organisation of the inspectorates in relation to the division of labour between its members.

According to country, the local level comprises districts or counties which constitute the area of practical competence of the labour inspectorate. In every case, the management of these districts or counties is the responsibility of one official. In Sweden - and this is the only country to mention such an exception - local management is by a government-appointed manager but assisted by a 'labour inspectorate board' composed of 10 members nominated equally by the social partners and who are in charge of monitoring or directing the activities of the inspectorate. A structure of this kind is only found in other countries (Denmark) at the highest level.

Taking into consideration the information available to us and which is not sufficiently detailed to enable a very precise typology to be outlined, the local methods of intervention may be characterised as follows:

- **Versatility.** With the exception of specific sectors mentioned in the introduction to this section (mining and quarrying etc.), in the majority of countries (Belgium, Denmark, Greece, Italy, Finland, France, Sweden, Portugal, Italy and the Netherlands) the inspectors have overall competence to inspect, in the companies under their jurisdiction, all the factors which affect health and safety. (The way in which the inspectors carry out their work in practice is examined in Chapter 3).

Within the inspection services, the hierarchic level, generally associated with the level of education rather than length of service, is also an element which determines the division of labour without entirely placing in question the principle of versatility. In France the inspectors look after large companies while the controllers, who have a diploma and a lower status, usually look after the small companies (which does not exclude a controller from accompanying an inspector on a visit to a large company). In fact, the responsibility for the work is such that everywhere, as the reports mention, a division of labour exists in accordance with the assumed complexity of the work to be done and also in accordance with the status the company occupies within the various grades. Naturally, a division of labour of this kind does not necessarily follow what is laid down in the book!

In this context, certain types of specialisation dominate. At the highest levels, inspection is closely modelled upon the law and on the application of the rules. The inspectors are primarily lawyers (France, Spain and Portugal). In other countries the inspectors are essentially technicians (engineers of various kinds) who have often worked professionally before joining the inspectorate. Those representing the lower grades are also very often technicians.

The idea of versatility is all the more pertinent here since, in the majority of cases, the control of the measures taken for protecting the health and safety of workers is only one part of the work of the inspectorate. In general, the inspectors are also responsible for promoting respect for all the laws which regulate labour and social security.

- **Sectorial specialisation.** In practically all of the countries there are still specialised inspectorates, even though these are progressively being united under the authority of the Ministry of Labour.[6] This is still the case in regard to agriculture in countries where this sector occupies an important place. It is also the case with mining and certain types of transport or activities which require a particular type of control. Not enough information on this subject was contained in the reports to enable any specific description of these to be made.

In the sector which may be called common law, the United Kingdom has set up an original structure to link their centralised and decentralised systems: an inspectorate which is mainly concerned with the industrial sector coming under the Health and Safety Executive whose inspectors are divided into field operation divisions, and a very decentralised structure whose inspectors come under the local authorities responsible for public health and the environment, who deal with the tertiary sector including services, recreation, public utilities and cultural activities. Whereas the first type fulfil a very classic function, the second has additional competence in the monitoring of food hygiene.

- **Versatile medical teams.** In reforming its public health system in 1978, Italy set up decentralised structures to be responsible both for workplaces and dwellings. Since that time this country has had an inspection service which is

---

[6] The Ministry of Health in Italy

strictly specialised in health and safety, the labour inspectorate having charge of all the other fields of activity usually assigned to it, i.e. the monitoring of work contracts, employment, working hours etc.

The monitoring services, which are in fact called 'prevention sectors', are regionalised. They are made up of versatile teams, half of whose members are doctors or nurses and the other half technicians. The object of these prevention sectors is to monitor the implementation of the law on health and safety and to develop preventive measures by drawing up global or sectorial plans within the companies. As in other countries, these inspectors may take immediate protective measures. In the exercise of their functions they are endowed with law enforcement powers.

- **The dual system.** Germany also has an original inspection system in the sense that the two inspectorates have, in principle, the same competences.

The labour inspectorate, in the classic meaning of the term, comes under the Trade Supervisory Authorities which are themselves attached to the Länder. The inspectors are responsible for monitoring the implementation of the laws, including those governing health and safety. Their role is polyvalent and their competence territorial. In their turn, the Berufsgenossenschaften (mutual indemnity associations) which, as the name suggests, manage the benefits for industrial accidents and occupational diseases and which are co-managed by the employers and workers, have a control and inspection organisation, organised on a sectorial basis: 20 for agriculture, 36 for industry and commerce and 51 for the public sector. Theoretically, all companies could be inspected by the two inspectorates whose purpose is now redundant. Since 1968, these two services have had to cooperate and, in practical terms, one may observe, according to the rapporteur, a division of roles between the two types of inspectorate: the agents of the Berufsgenossenschaften - who are more specialised - take charge of the highly technical inspections while the labour inspectors undertake more general inspections of health and safety.

To a lesser extent this dual system exists in some other countries with a similar division of labour. This is particularly the case in France where the agents of the CRAM also have the power to monitor and inspect.

In all the countries the inspectors have roughly identical powers. They are free to choose the place and time of their inspection. They may intervene immediately to bring an end to dangerous situations. They have powers of sanction whose application may in fact vary from one country to another: complaints against a negligent employer may be passed on, via the hierarchic bodies, directly to the judicial authorities or to the police for prosecution (Denmark). The number of sanctions imposed annually shows that the inspectorate still has every reason to exist.

2.1.2   Covering the risks

It is probably through the structures set up to cover the risks incurred by workers at their workplace that one can best understand the importance of history in forming an awareness of the problems of health and safety.

Today, in the 13 countries surveyed, all workers are covered against industrial risks.

An attempt may be made to classify the structures set up in three broad categories:

- **A specialised and co-managed public system.** In this case, the insurance schemes against industrial risks and occupational diseases are managed under the aegis of the competent ministry. The contribution rates are generally fixed by industrial sector and calculated on the basis of the risks incurred in that sector. In some countries, in the case of large companies, the premiums may be calculated as a function of the actual rate of accidents and diseases which occur. In the majority of cases, this system is co-managed by the employers and workers under the aegis of the State. This is the system chosen by Germany, France, Italy, Sweden, the Netherlands - and Belgium in respect of occupational diseases. The Belgium report is the only one to mention a different regime for occupational diseases and industrial accidents. It appears, however, that this is not the only European country where such a system is in force.

The case of Spain is very particular, insofar as risk coverage is managed by employer associations ("Mutuas") under State control. Dual control of these associations by both employers and unions has been introduced in 1995 but has yet to be put into practice.

- **Approved private insurance.** This second model may be found in Finland and Belgium in respect of industrial accidents.

The employer's obligation to insure workers may be fulfilled by private insurance cover of his choice on condition that it is approved. In Belgium such insurance is approved and controlled by the Fonds des Accidents de Travail (Industrial Accident Fund) whose management is joint, and in Finland by the Ministry of Labour which grants approval to the 20 companies concerned.

This system does not differ very much from the preceding one. The scale of premiums is fixed in the same way: a rate per sector for small companies where the risks are mutualised and pro rata for large companies according to the accident and illness rates.

- **Private insurance.** In the third model the employer may fulfil his obligation through insuring with the company of his choice without the necessity for the latter to have prior approval. This is the system in force in Ireland and Portugal. Private insurance companies may insure a lump sum in compensation but they may also offer full compensation based on public liability as in the United Kingdom. For the past ten years, Greece has also had a sort of mixed system. Companies may contribute to the general system of social security or go to private companies to insure all their risks including accidents and occupational diseases. Here, market forces determine the rates.

Although there are three types of system, the real division is between the two first mentioned and the third. This has to do with the accent placed, in the case of the first two, on prevention whereas the private insurers almost exclusively play the role of compensators.

This accent on prevention is translated into the making of rules, the financing or bodies for research and prevention by means of increased contributions, and by intensive counselling and educational activities by these bodies such as one will not find, or only very marginally, where it is a matter of private insurance companies.

Apart from this simple description, should the accent on prevention not also be imputed to the tripartite management?

2.2 Internal company structures

Two types of structures exist within companies for ensuring that the work done does not affect the health of the workers. Apart from the employer's own responsibility acknowledged by all the countries, health and safety have been placed in charge of the worker representatives and medical officers. The cost of operating these services means that they may not be made compulsory for small businesses where they are carried out by the staff representatives.

2.2.1 *Health and Safety committees and representatives*

The participation of employees in matters of health and safety is without doubt one of the best indicators of the attention given to these problems.

The election by the workers, or their nomination by the unions, of delegates for health and safety is a very old practice in countries with a long industrial history while this has only come into place since 1994 in Portugal. The ways in which these delegates are organised, and the thresholds above which they are compulsory, on the contrary, vary significantly from country to country.

• **The thresholds and methods of election**

The major problem in this area is that of the thresholds which results, in a number of countries, in employees in very small businesses being excluded from opportunities to take action and from the controls on health and safety. In any case it is necessary to distinguish in these thresholds between those which are concerned with the existence of one delegate and those involving the operation of a special committee.

– **a) Committees**

In some countries, there are no special committees, their role being fulfilled by the works councils. Where they exist, the thresholds vary from 20 to 50. From the information available, the following list has been drawn up:

> **Belgium:** This is the only country in which the health and safety committees are also responsible for the improvement of the workplace.
> These committees are compulsory once the threshold of 50 employees has been crossed.

The members are appointed from among the members elected during the works elections which take place every four years and their number is a function of company size. The composition of the committee is on a parity basis and consists of an equal number of representatives of the workers and of the employer.

Denmark: Unlike the elections to the works council, any employee may be a candidate but must be approved by the union upon election. Once elected, and as long as the company has over 20 employees, the delegates select from their number those who will form a health and safety committee. The employer in turn nominates an equal number to represent him.

The Netherlands: The Workers' Council acts as a Health and Safety committee. At present it is compulsory for companies with 35 or more employees to set up a Workers' Council. Companies with less than 35 employees are not compelled to do so, but the law regulates a limited participation on the part of all employees. There are no regulations on employees participation for companies with less than 10 employees.

Spain: Since 1971, health and safety committees are compulsory in companies with 100 employees. The new Work Environment Act (1996) has brought the threshold down to 50 employees.

Finland: Committees exist in all companies employer over 20 workers. Half of the committee is composed of manual worker representatives, one quarter of the members are salaried staff and one quarter represents the employer.

France: Like Belgium, the members of the HSC are elected by an electoral body composed of the members of the works council and delegates of the employees. The committee is compulsory for all companies of over 50 employees including temporary workers or those on fixed-term contracts.

United Kingdom: It is the unions, provided they are recognised by the employers, who nominate the delegates whose competence extends only to the category of employees they represent (except in the case of offshore sites). It is the delegates who can demand the creation of a committee. The surveys emphasise that these committees exist in almost 40% of the companies employing more than 25 people.

Greece: Since 1985, all companies of over 50 employees must have a works council.

Ireland: Since 1955 the creation of these committees has been left to the free will of the parties themselves. Since 1980, they have been compulsory in companies with over 20 employees.

Portugal: An agreement exists, in the construction, mining and quarrying sectors, which provides for health and safety committees. The 1994 Act which makes them compulsory has just come into force.

Sweden: Health and Safety committees are compulsory in companies with over 50 employees. In these committees, the unions have one representative more than the employer.

- b) Delegates

The committees which bring together the delegates and representatives of management in a collective body where, at regular intervals, problems of health and safety are discussed, only exist in companies of a certain size. Numerous countries have been seeking solutions which would allow the employees of small or very small companies to confer on one of their number this special role of vigilance. Failing detailed and exact information, we may mention here the countries whose rapporteurs have supplied some data.

In **Germany**, all companies, according to the report, have been obliged since 1973 to appoint a doctor and a person in charge of safety. Once these two persons are nominated, one or more safety delegates are then elected. The rapporteur does not mention any threshold effect.

In **Denmark**, companies with less than 20 employees must also have a delegate. Unlike companies with over 20 employees, the latter may be selected on an informal basis without recourse to elections.

In **France**, it is the staff delegates who fulfil this function in companies with between 10 and 50 employees. There is no provision for companies with under 10 employees.

In the **United Kingdom**, the presence of a delegate depends on his/her nomination by a union, regardless of the size of the company.

In **Ireland**, above the threshold of 20 persons, companies and their employees may appoint one safety delegate but this is left to their discretion.

Sweden is the country where, it appears, the presence of the delegates is best organised. In companies employing between 5 and 50 people, the presence of one delegate is obligatory. In the case of very small companies, **delegates to the regional safety committee** are nominated by the unions. The cost is borne by the new Swedish institute of work life and occupational health research. There are at present over 1,400 regional delegates. (It seems that regional delegates also exist in Spain but this was not mentioned in the report.)

* **Coverage**

Apart from the executives who formalise the organisation of this function, the few results of surveys which are sometimes supplied in the reports show that even though these bodies are compulsory they do not yet exist everywhere. In countries with a long industrial tradition these committees and these delegates are present almost everywhere in companies in the secondary sector of a certain size, a size which is much greater than that required by the book. The tertiary, private sector is less well covered since the risks here probably seem less serious.

Practically all the reports also mention exceptions. Some involve the high-risk sectors. Thus, in Portugal, before the 1994 Act, an agreement was made in the construction sector to initiate special measures in relation to health and safety. In Germany, France and the United Kingdom the rapporteurs mention special, reinforced structures for all the industries with 'special risks', the construction and public works sectors being in the front line of these, but also the fishing and nuclear industries.

Other exceptions concern State employees. In most of the countries concerned in the survey the rapporteurs emphasise how, in matters of health and safety, the State employer is not exemplary. Special representative bodies in this field, where they exist, have been set up at a late date. In the public service, industrial medicine is underdeveloped and those responsible for safety often non-existent. The alignment of the structures with those of the law should constitute an unquestionable advance.

2.2.2   Prevention services

The organisation of the prevention services is relatively diversified according to country. In this field, the length of time this has taken as well as some idea of the concept of public health should be taken into account in order to understand the differences found.

Within the prevention services two relatively different models are dominant although they cannot be compared word for word since their common denominator is, after all, the logic of prevention.

The first model that is found mainly in the Latin countries makes the doctor the central and unique pivot of the service. The typical example which illustrates this is that of France where the doctor is encouraged to collaborate with other experts (ergonomists, psychologists, sociologists etc.) in the creation of interdisciplinary activities. On the other hand, the prevention services such as those in Ireland or the United Kingdom may not even include any doctor even if, in practice, at least one is always found in large companies. In the Netherlands, the 1994 Act lays down the nature of competence for the four types of experts on the Occupational Health Services: an occupational health officer, a safety expert, an occupational hygienist and a work & organization specialist. Elsewhere, in Finland, Sweden and Denmark, generally speaking, alongside the doctors one finds specialists in the human and social sciences.

* **Freedom of choice of doctor and other prevention centre experts**

The first criterion common to all the countries is the freedom of choice allowed to the employer of the doctor or other experts who will work in the prevention centres. France and Italy require the doctor to be a specialist in industrial medicine. In all the other countries this specification is not obligatory. In some countries, finally, the choice of the doctor or the selected structure for prevention must have the agreement of the Health and Safety Committee. In all the others it is kept informed of this.

- **Company size**

The data contained in the reports is confusing since, in them, the size above which a company must employ a full-time medical officer or have a prevention service has been combined with the size when it is necessary to organise medical surveillance for its employees. Only very large companies are expected to have one or more full-time medical officers (e.g. those employing from 1350 to 3000 according to the sectors in Germany). In the case of some countries like Sweden, Denmark, Finland, Germany or France no threshold is mentioned, which probably means that all the employees - regardless of the company's size - are under medical surveillance. In the case of the others, on the other hand, the thresholds mentioned show fairly serious discrepancies: 50 employees in Belgium (HSC thresholds), 150 employees in Greece, 100 in Spain and 200 in Portugal, a threshold which may be lowered if the company's activities present particular risks. The time which the medical officers must devote to the company's employees is calculated pro rata based on their number. In the Netherlands there is no threshold; all employers are obliged to be assisted by an Occupational Health Service Unit.

- **How the prevention services operate**

Whatever the company size, it may engage the services of a prevention service or the doctor of its choice either full-time or part-time according to the numbers employed. Quite often, companies combine to make use of an intercompany service, whether this is designed upon professional or interprofessional bases (cf. the figures for Belgium and France where the centres must have the approval of the competent ministry). In Sweden, for example, in 1992, 900 private medical centres were reported, employing 10,000 doctors, nurses and psychologists. In Finland, where there are also intercompany services, the legislation prescribes that municipalities must provide such a service to any company which requests it.

The Netherlands set up, in 1994, an original structure which consists of creating health and safety services which are obliged to engage four experts: a doctor, a safety expert, a hygiene expert and a work planning expert. These centres are then certified if, in addition to the qualified experts, they meet the ISO 9000 standards. The quality of these centres is also tested with their clients. The service is paid for by the employer.

In the United Kingdom, as in Ireland, the employer may organise the prevention service as he understands it and is not forced to use the services of a doctor.

* **Financing the prevention services and the coverage of employees**

The responsibility for financing the medical surveillance of workers and of operating prevention services reposes entirely with the employer. Sweden, which until 1993 was the only country to escape this rule, has since that date reverted to common law. This has also had an impact on the prevention services whose number has gone in one year from 900 to 600. In Finland, in the case of small and medium-sized businesses, 50% of the cost may be reimbursed by the State if the prevention services used by the company meet a number of specific criteria.

Not all the rapporteurs supplied figures which would allow us to obtain a detailed impression of the working population covered by these services. The lowest figures are for Portugal, with a coverage of 12.95% of the workforce and Spain with 15%; the highest are for Finland with a coverage of 85% and Sweden with 80%. These averages sometimes conceal great disparities either between industrial sectors or between regions. Thus the British rapporteur shows a better provision of OHS for employees in the public sector while the opposite is the case in France. In the latter country the data published by the Ministry of Labour shows regional differences which, however, have no connection with those found in Italy.

The question which remains open is that of the care provided by a prevention service to all those workers who are not in permanent employment. This question not having been raised in the national reports, it is not possible to answer it here but this is a question which, in itself, deserves special examination.

2.3 The research structures

The position given to research on the subject of health and safety varies greatly. The structures set up, however, are in response to principles which may be classified in three main groups.

• Funding

Two essential methods of funding research may be observed in the different European countries. These methods are not without their impact on the structures set up.

a) Deductions from company social welfare contributions
A percentage of the amounts paid by companies is deducted from the contributions made to cover risks (accidents and illness) to fund a body which is itself responsible for conducting the research (e.g. the National Institute for Research and Safety in France) or commissioned for a more or less limited period to finance research (e.g. the Swedish Working Life Fund). Financing is, of course, not exclusive; other methods of funding are possible, either from the public purse or from the companies themselves.

b)   Funding by taxation

In the majority of countries research is financed by public funds in several ways.   The salaries of research workers in universities are taken into account in the State budget.   The funding of research is directly managed by the Ministry of Labour or by the regional authorities.   Here again, this method of funding is not exclusive of other sources, including private contracts with the universities or specific fundings from the insurance bodies.

- **The structures**

The Universities

In all the countries the universities remain an essential place for the development of research and education.   Naturally, the research there is specialised according to the discipline even if, for example, - and this was the only country to mention it - the specialised institutes of public health in Spain pursue an active policy on health and safety in the workplace.   As a general rule, the task of the universities is to provide teaching and research at the same time and it is they who are responsible for educating the researchers of the future just as much as the different specialists who intervene in Health and Safety.

The specialist institutes

These exist in the majority of the European countries (except in Ireland and Belgium).   Some countries have several organs of this kind which specialise in specific fields.   Some examples of these are:

Sweden: The National Institute of Occupational Health (NIOH) specialises in what the author of this report calls the classic fields of health and safety, and the Swedish Institute for Worklife Research, which is also interested in questions of worker participation, industrial relations and work organisation. These institutes merged 1 July 1995 into a new institute of work life and occupational health research (Arbetslivsinstitutet) with more emphasis on work organisation, industrial relations and labour market issues.

The Netherlands: The TNO Prevention and Health specialises in research into working conditions, and the Dutch Working Environment Institute (NIA), which works in the same field as the TNO but whose activities are concentrated on intervention, education and the dissemination of information.

Finland: The State Technical Research Centre, specialising in health and safety, and the Institute of Occupational Health, covering the whole field of research in relation to working conditions.

Germany: The BAU Federal Agency for Industrial Safety and Accident Research has a dual role: the improvement in practice of working conditions and the task of developing scientific knowledge. The BAFAM or Federal Agency for Occupational Health conducts research in the field of industrial health.

Spain: The National Institute for Safety and Hygiene at Work (IWSHT) provides information and training and carries out research.

France: The INRS, the National Institute for Research and Safety, apart from its own research activities is the centre for the resources, information and education essential for the taking of health and safety measures. The ANACT, the National Agency for the Improvement of Working Conditions, in its turn, has more of a role in encouraging research and the dissemination of knowledge rather than in actual research.

The characteristics common to all these bodies are interesting:

- They bring researchers in different specialisations together in one organisation;
- Although they carry out fundamental research, they chiefly carry out a good deal of applied research;
- They play an essential role in disseminating the results, by means of training experts, union officials, management personnel etc. or in disseminating information and teaching materials accessible to the widest public (information booklets, films etc.)

• Coordinating the research

When this is done, it is mainly achieved by means of funding. Few countries mention the existence of specific funds (Sweden and Germany). In this case, this may be managed by a tripartite body which approves the programmes. As a rule it is the government agencies, particularly the Ministry of Labour or of Health and Social Security who play a deciding role in encouraging and coordinating by means of funding or by means of direct commissions for the bodies which they administer.

It should be emphasised that this coordination leaves a good deal of autonomy to the initiatives of the university centres. The social partners also intervene whenever there are tripartite bodies in existence to receive them. Community funds have played a deciding role in initiating the development of research in countries whose infrastructures in this field are still underdeveloped.

Finally, the unions in numerous European countries, have allocated the means for conducting their own research, either by creating their own structures or by relaying their specific requirements to the universities or public research laboratories.

CHAPTER 3

THE ACTORS

Knowing the actors is one possible key to understanding the efficacy of structures put in place and policies adopted.

Here once again, we cannot in any way claim that our presentation is exhaustive. The data on which it is based is taken exclusively from the national reports. This is a topic on which investigations conducted at national level can sometimes seem wanting. The responsibility for this lies, as we have already seen, more with the authors of the survey guide than with the rapporteurs.

Health and safety involve a great many actors whose status differs and whose remoteness from, or proximity to, the workplace varies in degree. The way in which they are selected, the training they receive and the roles assigned to them are good indicators of the way in which their structures operate.

3.1 The institutional actors

In all the countries of Europe, safeguarding health is at the root of the earliest laws governing labour. Naturally, the very concept of 'health' has developed considerably since these first labour laws which are now a century old and the legislation itself is a good index of the approach taken to this problem in different countries. No longer is it simply a matter of maintaining the health of employees, limiting their working hours, prohibiting the involvement of young people and children in certain kinds of work, or protecting employees from the most obvious hazards, attention now needs to be given, too, to the psychological dimension of the effect of work on the individual, as emphasised by the tendencies of the more 'advanced' countries in that direction. This development in the perception of health and safety is reflected in part in the development of the role and the status of the actors entrusted with application of the law.

3.1.1 The inspectors of labour

The uniformity of the language used conceals a great deal of diversity not so much in regard to their status as in methods of recruitment and roles played.

• The method of recruitment

The method of recruiting inspectors of labour varies considerably from one country to another. Some countries recruit their inspectors bureaucratically, within the strictest meaning of the word (mainly among the Latin countries, for example France, Spain and Portugal), operating a competition open to candidates who hold a third-level degree without specified specialisation, or a diploma equivalent to two years of university education for those intending to assist the inspectors or to carry out supervisory activities in small-scale establishments. The competition is adapted in level to the diploma concerned, thereby creating a hierarchy of grades. The content of the competition programme determines the level of knowledge required, which is generally highly legal in orientation.

In the Scandinavian countries, as also in Great Britain and Ireland, recruitment is more pragmatic by nature and more diversified. A competition is not the favoured way of selecting candidates. Rather than the legal mind, the kind of candidate sought is a technologist, namely an engineer, a chemist, a mechanical engineer, an architect or, at a lower level dependent on duties to be performed, those with vocational school qualifications. While not the norm, candidates are sometimes recruited from among holders of a specialised degree in health and safety (Sweden). Very often, vocational experience and professional knowledge of a given sector are stipulated for recruitment, which can be conducted either at national or at local level (in Sweden, for example, each district chooses its personnel as a function of its needs).

Whether recruitment is based on a competition or follows more traditional methods, the level of initial education determines the grade occupied by those who exercise this function.

The training which follows recruitment varies greatly not alone in duration, but also in form. The longest period of training would appear to be that offered in Great Britain (three years, comprising six-month theory followed by practical courses). In all cases, training is largely made up of practical courses or an apprenticeship. This training is given by specialised institutes, or less often by specialised departments of the universities.

In-house training is a means of keeping knowledge up to date which is employed in almost every country.

In the one particular case of Italy, inspection services are multidisciplinary services in which doctors are about 30% of the personnel; the remaining being engineers, safety technicians, nurses and administrators.

• Their status

No matter how they are recruited, inspectors of labour everywhere enjoy a status which gives them total independence from the companies they are called on to supervise. This independence is expressed throughout by a discretionary right of inspection and by a professional code of ethics which applies to all grades.

While the status of inspectors may be similar, their powers are not the same everywhere, even if they have a wide degree of latitude in exercising them. In Italy, though this is the one exception mentioned, the officials of the health service have policing powers. Everywhere else, serious offences liable to fines or legal penalties are prosecuted through the official channels. Several reports refer to the power conferred on the inspector of labour to suspend any operation deemed to be positively dangerous.

- **From versatility to expertise**

While their status may be similar, the specific role of an inspector, as might be expected from the diversity of recruiting systems, differs very considerably in individual countries. At the risk of taking a simplistic view, the attempt could be made to classify the roles of inspectors as a function of their expertise and their specialisation, these roles being very much dependent on national preferences in the organisation of inspectorate.

In a number of countries, the activities of the inspectorate of labour cover not only the problems of health and safety, but also a good many other aspects of labour legislation. The inspectors' powers are territorial and, within their own districts, they have authority over almost every enterprise (there may be some exceptions in the area of very specific activities such as mining, transport and agriculture). This structure is found in Spain, Portugal, France, Belgium and Greece.

In the above situation, inspectors do not specialise by activity or vocational sector; the supervision of health and safety represents a part, and only a part, of their role, which is mainly to supervise the application of the law. Which is not to say that their intervention is not in the process of development. Every country makes mention of the widening range of tasks entrusted to their inspectors, broadening to include incentives towards improvement in working conditions, advice and support for new ideas.

In some cases (France), extending the Inspectorate's role to employment and vocational training is leading, despite evidence that some inspectors are specialising in these areas, to a spectrum of skills so broad that often it can only be contained by restricting it to the limits of applying the law.

Other organisational structures, on the contrary, have chosen to govern the specialisation of their inspectors by criteria which combine specialised vocational and territorial authority to varying degrees.

In the United Kingdom, industry is supervised by the H.S.E. (Health and Safety Executive), while services are subject to inspectors attached to the Environmental Health Department of the local authorities whose duties can extend to controlling food hygiene.

In Ireland, supervisory specialisation is based on the skills of the officials. In Denmark, intervention is in keeping with specialisations. The same applies in Finland where, in addition to a more 'generalised' form of inspection, there is a growing body of inspectors who are expert in more specific problems.

The German situation, where a dual system is operated, is specific and original. In concrete terms, the principal features referred to above are also to be found in this system. The prevention of accidents is entrusted to the inspectors of the Berufsgenossenschaften (mutual indemnity associations), while the 'Trade Supervisory Officers' have more general terms of reference. According to our German rapporteur, there is a sharing of roles: an inspector (the Trade Supervisory Officer) plays the general-practitioner role, while his colleague from the Berufsgenossenschaften centres his intervention on his own expertise.

The philosophy behind inspection, its organisation and also its context play a key role in determining the way in which it may be conducted.

• Modified roles

However, all the reports unfailingly mention not alone the complexity of the inspector (or controller) of labour's duties, but also their changing nature.

The Finnish report provides the most vivid portrayal of this development: "the inspector is no longer a policeman, he is both counsellor and consultant". This development in the perceived role of inspection is the consequence of a modified perception of health and safety policies and of knowledge in this field. All the countries of Europe are unanimous in agreeing on the importance to be attached to prevention. Accordingly, inspectors are expected not alone to cease behaving solely as 'policemen', but to be capable of performing an 'external audit' (Belgium), able not so much to provide a solution, as to "describe a problem, discussing the hazards and the rules, and determining the priorities" (Denmark). In that country, the inspector sends a company 'improvement notes' to which the employer is obliged to reply, also 'information notes' whose intention it is to provide him with information on which action can be based. The word the reports most often use to describe the inspector's intervention is a 'systematic' or 'holistic' approach (Sweden, Belgium, Portugal). The inspector is expected, also, to demonstrate teaching and negotiating capabilities in communicating his message and convincing or encouraging the opposing parties to cooperate in this area. In Sweden, the inspector can conduct a 'systematic inspection' in the course of which he will examine the overall system put in place by business enterprises or authorities to prevent occupational injury and disease.

Such an evaluation of the way in which the profession of labour inspector is perceived encounters difficulties of various kinds.

• Insufficient manpower

It would appear, from the information available, that the inspector's burden is a very onerous one.

Table 2: Number of inspectors per number of employees

Belgium	1 for every	14,000
Denmark	1 for every	9,000
Finland	1 for every	5,000
France	1 for every	10,700
Ireland	1 for every	20,071
The Netherlands	1 for every	16,667
Portugal	1 for every	11,000
Sweden	1 for every	10,500

(The figures in Table 2 should be viewed with caution, in the perspective of individual company size)

In some cases, this burden can be difficult to reconcile with a broadening of the inspectorate function. Moreover, as mentioned in several reports (Greece, France, Denmark), the balance to be maintained between the functions of counselling and diagnosis is a delicate one and arouses a degree of unease within the profession. Environment is another aggravating factor, penalties being difficult to impose if the correction required is a threat to employment.

It should be said, in conclusion, that the characteristics sought in recruitment do not necessarily correspond to the activities now expected of an inspector. Two career types dominate the profession, namely the lawyer and the engineer (or in Italy, the doctor). Its creation was closely associated with the beginnings of industrialisation and mechanization. The growing development of the services sector is altering the traditional problems in the area of health and safety. Such is the new reality confronting the inspector of labour at the present time.

3.1.2 The Experts

To carry out their duties, inspectors in a number of countries can call on the aid of specialised experts who may either be attached directly to the Inspectorate of Labour (as in France, for example), or belong to ad hoc bodies (public research bodies, bodies associated with the various insurance systems). These experts, where mentioned in the reports, are consultant engineers or medical officers. In the Scandinavian countries in particular, they may also be ergonomic engineers, or specialists in the behavioural sciences and work organisation. The Finnish report specifies that the inspector can arrange to be accompanied in his duties by an expert of his choice.

However, the role of these experts cannot just be reduced to one of close association with the inspectors of labour.

These experts are often employed by the insurance institutions and can play a very important part in the area of prevention within the company. In some cases, as is especially true of Germany, their function is directed in the main towards preventive policies and the evaluation of company efforts to reduce hazard, which efforts are likely to have an influence on the level of insurance premiums (France, Spain).

But these experts are also placed at the disposal of business enterprises through the medium of public or semi-public bodies whose terms of reference are to assist in the 'modernisation' of companies, changing the way work is organised, making the workplace more ergonomic and introducing new technologies. These bodies engage experts from various disciplines or specialisations capable of helping companies to conduct a diagnosis dealing strictly with health and safety, or to include the latter in the broader concept of working conditions. Finally, as we will see in the section devoted to the subject, the research workers can themselves play this role of expert at company level.

3.1.3 *The prevention service professionals*

● **Medical officers**

The majority of the countries in question refer in their reports, sometimes confusing them, to two categories of medical officer, namely those acting as experts either within the labour inspectorates or the insurance schemes' controlling bodies, whether semi-private or private, and those who practise on a daily basis in business enterprises. These two categories of medical officer do not have the same status, their responsibilities are not the same, they do not play the same role and sometimes even their training is different.

We will only be dealing with the first group at this point, the role and function of company medical officers being covered in the second part of this chapter.

In all the reporting countries, consultant, researching or inspecting medical officers hold a specific specialist qualification obtained through training which, as a general rule, extends four years beyond obtaining a medical degree. Besides those who devote their activities to research in universities or specialist medical schools, the majority of these doctors exercise a controlling function for the insurance institutions, assessing degrees of disablement after an accident or an illness, diagnosing occupational diseases and following up their development, etc.

Italy is the only country in which medical officers are called on to exercise controlling functions, especially in the application of health and safety regulations. Under the latter heading, they are invested with criminal investigatory powers and their role is similar to that held elsewhere by the inspector of labour.

- **Other experts**

While other experts may be mentioned as part of the preventive structure, none of the reports has anything to say on the way in which they conduct their activities.

3.1.4 Research workers

A link with the role of the expert has already been suggested in the part played by the research worker. In the field which concerns us, we should recognise a continuum between basic research, applied research and expertise, rather than any great distinction between the two.

With the exception of Denmark, on which we have no data, European countries appear to differ greatly in terms of the number of research workers who devote their attention to the conditions and hazards of work.

An initial division may be seen between the less well-off countries whose industrial tradition is the most recent, namely Ireland and Greece, and the rest. In these two countries, Community aid plays a decisive part in the development of research.

A second division exists between basic and applied research. The data we have on this topic is relatively imprecise but, where available, it indicates that as a general rule, basic research in the area of health and safety mainly involves doctors and epidemiologists, but also chemists, biologists and representatives of all the scientific disciplines who can work towards a better understanding of the hazards employees incur through their working conditions and the handling of hazardous products or substances.

Over about the last fifteen years, a new body of research workers has appeared in this field: specialists in the social sciences, primarily ergonomists, then psychologists and sociologists. Except in Sweden and Finland, the latter two categories of research worker are very much in the minority in the area of health and safety, specialising rather more in problems associated with the organisation of work, job content and labour relations.

Lastly, research workers attached to the engineering sciences play a determining role in research into the safety of equipment.

A final point to be emphasised is the very small number of economists (indeed the non-existence of representatives of this discipline in the majority of public bodies) who devote any part of their research activities to health and safety.

Few reports mention the proportion of work explicitly devoted by research workers to the evaluation of measures put in place (the INRS in France), but this absence of evaluation is no doubt attributable to choice in the area of research policy, rather than the personal choice of the researchers.

The role of research workers is very closely linked to the research structures and policies adopted in individual countries. However, it would appear, from the data available to us and subject to further information, that the very high degree of specialisation required to reach a level of excellence in a particular discipline is not always favourable (except perhaps in Sweden and Finland, or to a lesser degree in Germany) to the creation of multi-disciplinary teams such as are more readily to be found at work on applied research.

In some countries, if they have a specialised body for the purpose, like the TNO in the Netherlands, the INRS in France and the Institutes of Occupational Health in Finland and Sweden, research workers with a specific background devote a good part of their time to the transfer of knowledge. Which means that along with a profound knowledge of their research work, they must have the ability to communicate the results in accessible terms, through publications or training, to experts and the relevant personnel at company level. This role and these terms of reference, both essential, ought to be developed in future.

3.2 The actors at business enterprise level

The quality and effectiveness of the health and safety policies adopted reveal themselves at company level. The quality of research and the extent of the controlling measures cannot replace a willingness on the part of the actors *in situ* who are especially entrusted with these matters in individual companies.

No logical pattern has been followed in presenting these actors, who cannot in any case be seen in any order of importance since each of them has a determining role in this context. Here more than anywhere else, the quality of action taken depends far more on coordination between the different actors and on their ability to form a genuine team.

3.2.1 Company medical officers and nurses

In all the countries, companies are expected to watch over the health of their employees by placing health-care staff at their disposal. These obligations vary with size of company, only the largest being required to provide a permanently present medical service.

● **Company medical officers**
Company medical officers, whose terms of reference are above all preventive, carry out their duties within the framework of the legislation which provides for their intervention.

● **The training** of these medical officers is tending towards harmonisation among all the countries of the EU.

However, specialist training in industrial medicine as offered by the medical faculties still differs a good deal from one country to another. Such training is almost non-existent in Greece, where none of the universities has a department of industrial medicine. The course in Belgium is for one year, which is to be extended to four years; in Germany, it takes two years of specialist training to become a company medical officer, while four years are required to obtain the qualification of 'Physician in industrial medicine'.

In Italy, the inspecting medical officers are required to hold a certificate in industrial medicine which is likewise acquired after four years of specialised study.

In France and the Netherlands, all doctors wishing to practice within a company should be holders of a specialist certificate (requiring four years' study). In the latter country - the only such case in the sampling - the doctor is required not alone to work towards prevention, he must also verify absence on medical grounds. As a general rule, company medical officers in the Scandinavian countries have a specialist qualification in industrial medicine, but this does not seem to be obligatory.

Accordingly, company medical officers have attained the level of general practitioner as a minimum on which they can build through permanent specialist training.

• The principal **role** of the company medical officer, whether practised within a particular company or in an inter-company group, is preventive. This is a word whose meanings may vary in practice. Primarily, it involves supervision, at intervals of varying regularity, of the health of employees, seen far more from the individual's point of view than in the collective context of work in general.

Next comes verification of an employee's fitness to hold any particularly difficult or dangerous post, or assurance of his satisfactory reintroduction following absence due to illness.

Going by what has already happened in Sweden and Finland, we are also witnessing a development in the perception of the role of the medical officer, who is now expected to intervene more directly in an active policy of prevention within the company. The law in France provides that the medical officer should devote one third of his time in the company to the examination of workplaces and that he should be regularly consulted on the introduction of new technology. He attends meetings of the CHSCT (Health, Safety and Working Conditions Committee), presents an annual plan of activities in larger companies or sets up hazards files in companies with from ten to fifty employees. More recently and provided there is a company agreement to the effect, the medical officer can carry out his medical inspection once every two years, on condition that he spends the time thus liberated on the analysis of working conditions within (in most cases) a multi-disciplinary team.

In contrast with the inspectors of labour, whose status is relatively similar throughout Europe, the specific status of a company medical officer is highly variable. They are always paid by the company, but their answerability to the employer differs according to whether they are members of the company's staff, or of an inter-company service centre. Practice of the profession and specific skills differ very considerably according to whether the medical officer is working full-time for the company or providing his services as a second job.

Among all forms of specialisation, industrial medicine still enjoys little prestige in a good many countries. The reforms achieved or in progress (Belgium) aim at introducing the same requirements for this specialism as apply in other medical disciplines. This specialist training is non-existent in Greece, and not very widespread in Portugal or Ireland.

● **Nurses**

While the training of medical officers is improving and while there may be a foreseeable long-term requirement for specialisation by all company medical officers, the same cannot be said for nursing staff. With some exceptions (Sweden, Finland, Denmark, the UK, Ireland and the Netherlands), nurses receive no extra training with a view to pursuing their career in industry. Undoubtedly, there is room for development in this area too, given that this category of nursing staff is often present on company premises on a more continuous basis than the medical officer. Besides her duties in assisting the medical officer, it is also the nurse who meets and looks after employees who call on her in connection with minor illnesses or minor injuries.

3.2.2 Safety officers

Only a few reports refer to safety officers, which, of course, does not suggest their non-existence in the countries who make no mention of them.

In all the countries, the person finally responsible for health and safety is the head of the company. He it is who must put in place all the facilities for which the law provides. He is responsible when any occupational injury or illness occurs. But keeping to the law is not enough to prevent injury. A great many factors contribute to the quality of working conditions, including observance of strict standards in the handling of toxic or hazardous products or in the safety and proper condition of machinery employed, but also involving ergonomic workplace design, genuine cooperation with employees and their representatives in these matters, work planning which avoids monotony and repetitive tasks, a pleasant atmosphere at work, etc.

As a general rule, the heads of companies, whether large or small, have no specific training in carrying out these duties. Employers' associations may offer training courses on this subject, or organise meetings which allow an exchange of information or experience (as in Germany, Denmark and Finland, for example). Public or semi-public bodies may also offer specific training of this kind. The knowledge a company head may have with regard to performing this duty is often far from commensurate with his responsibilities in the matter.

In larger companies, safety is often identified as a specific function which can occupy a whole staff of people with different skills. Some countries specify the duties of this function and the kind of skills its holder should have. In Germany, for instance, the 'safety officer' should be an engineer, a 'master craftsperson' or a technician who should not only have practical experience, but must also have attended a recognised training course if he is to perform this function. In Belgium, the person in charge of safety must attend a training course whose level depends on the nature of hazard or risk encountered (level 2 for moderate risk, level 1 for serious risk). In France and the United Kingdom, the holder of this office should be competent, but it is left to the company's discretion to determine the criteria governing this competence. In the Netherlands, safety officers are experts within the context of the law concerning the working environment. The report from that country indicates that there are about 1,400 such officers, but does not specify the criteria by which they are selected. Large companies in the industrial sector generally have well fleshed-out safety departments employing experts with different skills but predominantly engineers and ergonomists, who are joined by the company medical officer. Such departments are especially well developed where the use of particular machinery or products constitutes a source of danger.

The report from the Netherlands explicitly mentions the decisive role played by management, and supervisory staff in particular, in the area of health and safety. The same comment would no doubt apply to every country. But it should be said that even if sensitised to these matters, supervisory staff often lack the training required in practice. However, there are some notable exceptions which deserve to be under-lined. In Sweden, for instance, all 'supervisors' are required to attend a course entitled 'Better Work Environ-ment'.

3.2.3 Health and safety representatives

Health and safety representatives have a very decisive role to play. While long-established in the older industrialised countries, they are only beginning to find their place in countries like Greece, Portugal, Italy or Ireland.

Although not always the case, health and safety representa-tives very often issue from the ranks of the trade unions, either because directly nominated by the unions, as in the United Kingdom, or because they need to obtain the recognition of a trade union organisation once elected (Denmark). In countries in which trade unionism is in severe decline, as in France, the link between the members of the CHSCT (Health, Safety and Working Conditions Committee) and the unions is becoming ever more tenuous.

The intervention of these representatives is of two kinds.

Firstly, they act **collectively**, in a committee which is differently comprised in different countries but whose principal characteristic is to bring together representatives and employers to deal exclusively with problems in the area of health and safety, on a regular basis (from the information in our possession, these committees meet about once every three months).

This collective activity is very varied and especially important when it comes to examining changes to the working process or the introduction of new technology. The representatives have a particular role to play in connection with injuries at work and occupational disease. They can hear the inspector of labour and accompany him on his inspections. In France, the only country to mention this specific feature, the committee can arrange to be assisted by an expert in connection with serious hazards or projects which change the conditions of health and safety. In Denmark, all of a company's departments should have a 'safety group' comprised of two members, the representative and the 'supervisor'.

It should finally be noted that these committees play an extremely important part in supervising the training and informing of employees in and on health and safety.

However, the representatives also act on an **individual** basis, especially in small companies where they fulfil the function of the works council in the matter of health and safety.

In larger companies, the representative's function is not, of course, limited to attendance at committee meetings. He operates on a daily basis and must give constant attention to health and safety problems. Coming from the ranks of the employees, the representatives can play an essential teaching role in this field. In particular, their knowledge of problems at work is specific and independent. To allow them perform their duties, they receive training which varies considerably in duration from one country to the next. This training seems to be longest in Sweden, where each representative is required to attend a basic course of 40 hours within the six months following his election. The training period in Denmark is 32 hours; in France, it is five days (about 35 hours) for representatives in large companies (more than 300 employees) or three days (about 20 hours) for those in companies with fewer than 300 employees. These training courses are generally paid for by the employer.

The training is given by institutions approved by the relevant Ministries of Labour or by specialised bodies dependent on that ministry, but may also be the responsibility of the trade unions, as in the United Kingdom where they benefit from state aid for such training.

Even when training is not prescribed, all the countries offer various forms of training specifically intended for these representatives.

Over and above obligatory training, unions everywhere accord particular attention to these actors and provide extra training sessions at regular intervals.

3.2.4 Employees

While it is for employees that the relevant facilities are put in place, they also have an essential actor's role to play in the level of attention paid to health and safety problems. A number of countries make provision for explicit employee information and training on a particular workplace, the dangers presented by a given machine, by use of a particular product, the precautions to be taken in the handling of instruments, the obligation to wear personal protective equipment, etc. The employees' vigilance in these matters depends to a very large degree on the quality of the information and training they are given. French legislation, for instance, stipulates that the employee should be trained in an 'appropriate' and 'practical' manner and that this training should be repeated. This specific training of employees is provided in Belgium, Finland, Germany, the United Kingdom and Sweden and is to be introduced gradually in countries so far without such measures.

Employees pay all the more attention to health and safety problems if they have been sensitised to them while at school and from a very early age. Sweden appears to be advanced in this respect, having installed safety representatives in its secondary schools. This has been done in France in recent times, in the secondary schools for vocational education. Wherever it takes place, sensitisation and training begins with health, hygiene and nutrition before approaching the problems of environment. Where it exists, training in health and safety is given in the main by the technical schools and the engineering schools. But it should also be noted that greater attention is now being paid to these problems by national and local authorities with responsibility for training.

CHAPTER 4

THE POLICIES

This chapter covers only a few aspects of the policies introduced to improve health and safety at the workplace, mainly the national ones. The national rapporteurs were not in fact asked to investigate policies adopted by the social partners, except in passing, or the initiatives taken by companies themselves, as it would have required work on quite a different scale to that envisaged by the Foundation.

4.1 From supervision to prevention

While, as we have already said on several occasions above, prevention is assuming an ever more important place in the policies adopted in the various countries, supervision and the penalties associated with the investigation of offences remains an essential part of national policy.

4.1.1 *The cost of industrial injury and occupational disease*
Only half the national reports refer to the results of work on this topic, there being several reasons for these omissions.

Economic calculations require the existence of a very substantial basis of data, which is still far from being the general situation. It is also difficult, when examining data on absenteeism, to distinguish information on industrial injuries and occupational diseases which are directly associated with working conditions.

There is likewise a need for economists to take an interest in this kind of research, or to encourage them to do so. Among all the research institutes, whether established expressly to study health and safety, or part of broader university groupings, economics is the discipline with the least representation.

Results, where to hand, are very mixed, the British and
Italian reports being the most fully-developed. In the United
Kingdom, a research project conducted under the aegis of the
HSE estimates the consequential cost to companies of industrial injury and occupational disease to be between £11 and
£16 billion per year, which corresponds on average to between
£170 and £360 per employee. The cost to the economy is
estimated to vary between £11 and £16 billion, or 2 to 3% of
GNP. In Italy, detailed research into industrial injury alone
has estimated that it cost 3.05% of GNP in 1991. In Germany,
injury alone cost 18 million ECU to the community. In
Finland, the figures presented, without stating how they were
calculated, indicate a band of between 5 and 15% of GNP. The
Finnish report is the only one to mention research aimed at
designing software to allow analysis of these costs at
individual company level. Other reports provide incomplete
figures which, apart from their imprecision, combine with
those indicated above to underline **the importance of the
economic consequences of industrial injury and occupational
disease, both for business enterprise and for society.**

Awareness of these costs naturally induces the authorities to
accord even greater importance to prevention and supervision.
However, we cannot be certain that such analyses are enough to
direct individual decision-makers who must judge between the
immediate cost of investment or reorganisation, and hypotheti-
cal benefits whose costs are largely mutualised.

4.1.2 *The inspection and supervision of health and safety*

While the structures, roles and training of those entrusted
with inspection are still quite different from one country to
another, there is significant convergence in the direction of
policy in this domain.

Within their supervisory terms of reference which naturally
remain in place, the inspection services have responded to
criteria which sometimes gave them too strictly legalistic or

routine-minded a vision of their work by profoundly changing their activities. Allowing for the means at their disposal, they no longer develop their activities on the basis of random inspections alone, but more systematically, taking priority action either in certain industries like construction, chemicals, etc., or on specific nuisances such as noise, the use of particular toxic products, etc. They also utilise the whole range of instruments available to them, from recommendations, through injunctions, to penalisation.

To the greatest extent possible, prevention is taking precedence over punishment. The role of the inspectorate is becoming an advisory one capable of auditing (Belgium), issuing an 'improvement note' based on a description of the problem, a presentation of the hazards incurred, the rules in force and solutions which should be applied (Denmark). Its role may also be to advise and encourage the various parties to work together, developing a broader vision of health and safety which will encompass not only the physical or material factors, but also the organisation of work and the workforce.

This reorganisation, referred to in all the reports, is not without its problems and has caused discontent within some inspectorates (as indicated in the French report). The discontent is not attributable to reluctance in face of the new approach, but to the means required to put it into practice (manpower in particular) and to the extent of knowledge and know-how needed to perform such an office.
As emphasised in the Portuguese report, the inspector would have to be an expert with an overall view and at the same time a specialist in human relations and negotiation, gifted with great powers of persuasion.

Following the example of what is already in place elsewhere, the reorganisation of some inspectorates (France and the UK) aims to give them more effective means as a group, but the economic climate is putting a brake on the likelihood of providing means compatible with advertised ambitions.

4.1.3 Penalties

A supervisory policy only makes sense if the offences discovered are punished.

The inspectors of labour have independent authority to identify and assess an offence, it being entirely their decision, in the area of prevention, to record it and have proceedings initiated through the official channels. The same does not apply to injuries, of course, in which case an enquiry and proceedings are automatic.

There are two types of penalty, which may be concurrent.

• Financial penalties

As a general rule, the authorities establish a scale of fines proportionate to the type of offence. The average fines imposed in Germany may be cited to illustrate this point: 5201 ECU for failure to observe official working hours; 40,816 ECU for non-observance of technical rules relating to health and safety; 52,016 ECU in connection with the working environment. The inspectors of the MIAS (mutual insurance) can impose fines of up to 10,403 ECU for violation of the rules on prevention of injury.

Just as in Germany, offences in France, Sweden, Belgium and Finland are graded on a scale corresponding to their seriousness. This scale may include minimum and maximum levels. The fine for certain offences (France, Belgium) is calculated on a per capita basis, being multiplied by the number of employees exposed to risk.

In Britain and Italy, the level of these fines has just been considerably increased to render them more dissuasive.

Sweden seems to be the only country in which the State as employer is treated like any other head of enterprise (at least theirs is the only report giving this indication), meaning that it may be ordered to pay a fine.

• Penalties

The inspectorate bodies have authority to propose penalties, which are then communicated to the competent legal authorities through the official channels or through the police (Denmark). The courts charged with enforcement of the penalties can adjust the proposed sanctions by including economic factors in their judgments, which generally results in maximum penalties being the exception rather than the rule (which the labour inspectors in France, for example, deplore). In Sweden, penalties imposed by an inspector give rise to an appeal procedure within the inspectorate. Some reports mention the fact that, even in the area of prevention, certain offences may result in penalties which are not merely financial and therefore in the last analysis borne by the company, but also punitive, which means they may be recognised as the personal responsibility of the employer. In Italy, health and safety penalties are automatically recorded on the employer's police record. In connection with certain specific offences (absence of an injuries prevention plan) or with subsequent offences, French legislation provides for imprisonment or prohibition to hold the post of company head for a prolonged period of time (5 years).

As emphasised by the British report, there is no certainty of the effectiveness of financial or even punitive sanctions in the area of prevention, even if, as mentioned in several reports, the number of penalties increases while their levels remain low. To serve as a proper deterrent, fines must be high. But where they are too high they lose credibility if their application places the survival of the company in jeopardy, a risk no adjudicator is going to take in the context of an unemployment crisis. Even where prescribed by legislation, a prison sentence is imposed only in exceptional cases.

Despite these reservations and irrespective of penalties, the supervisory policies have been highly effective. They can no doubt be considered responsible, as they are in the Danish report, for the reduction observed in industrial injuries and occupational diseases. The combined effect of supervision and more active preventive policies should contribute to an improvement in the results already obtained.

4.1.4 Policies in the area of insurance

As we observed when describing these organisations, the choice of insurance system to cover risk in the matter of occupational injury or disease is not indifferent to the way in which a preventive policy might develop. Such a policy may take two distinct forms, namely an active system of prevention and compensation operated by the insurer, or a financially based policy adjusting premiums as a function of the results obtained.

• An active policy of prevention and compensation

The involvement or non-involvement of the social partners and the State in the management of insurance schemes would seem to be the key factor explaining whether they invest in a preventive rather than a merely compensatory policy.

Compared with the private insurance companies, who limit their action to increasing premiums in keeping with increased risk, mixed schemes or those integrated with the social welfare system, either directly or through *ad hoc* organisations which they finance in full or in part, are developing strong incentive policies in the area of prevention.

Two examples may be taken to illustrate the foregoing. Firstly, the promotion, in Germany, of **'Health Circles'** which operate on the lines of quality circles, simultaneously mobilising and training employees in general problems of health and safety. The second example would be the **'target agreements'** contracted between the Caisses d'Assurance Maladie (health insurance funds) and companies or industries to encourage the implementation of more effective preventive policies. This insurance system gives financial aid to companies, especially SMEs, to invest in preventive equipment, in the form of advances which can be changed to a rebate on premiums.

But not only is prevention is better supported by this form of organisation than by the private sector, the same may also be said of **the quality of compensation.**

Here again, the German insurance system is undoubtedly a good example, having its own clinics recognised by the hospitals, or clinics specialising in the treatment of industrial injury or occupational disease.

Finally, it should be said that the countries which, in recording causes of absenteeism, fail to distinguish those strictly associated with hazards at work (as seems to be the case in Sweden) are affected by the crisis touching all of Europe's welfare systems. The financial imbalance affecting these schemes leads governments to introduce, simultaneously, restrictive policies in the area of employee compensation and increases in employers' and employees' contributions (the Netherlands, Portugal).

Generally speaking, the reports make no mention of the level of financial compensation. However, the Irish report mentions the very high cost of legal proceedings in a system dominated by private insurance which appear to be undertaken almost automatically in the event of occupational injury or disease and which result, without doubt, in a variety of perverse effects.

• An incentive policy based on adjustment of premiums

Almost all the reports mention the existence of an adjustment of premiums as a function of a company's record in the area of occupational injuries or disease (Germany, Finland, France, Ireland, United Kingdom, Portugal). Few reports give details on the overall amount of premiums and adjustments. Where the amount is indicated, it makes an impression: 8,119 billion ECU in Germany and 6,902 billion ECU in France. The German report also gives the sum total of 'bonuses' allowed, being 1.35% of total premiums in a year, or 109.9 million ECU, while 'surcharges' represent 4.32% of insurance premiums or 350.86 million ECU.

The 'bonus' and 'surcharge' (upgrading) system is also operated by the private insurance companies, but doubtless less insistently in a competitive environment. The Greek report mentions the theoretical existence of such a system and its non-application in practice, while the Irish report calls to mind the very high cost of insurance associated with legal proceedings undertaken in all cases of compensation, leading some employers to cancel their policies and to suggestions for controlling the levels of compensation.

Mention should also be made of an original option offered to larger companies in Finland. These companies are not obliged to insure themselves in so far as they are capable of assuming the full consequences of occupational injuries and disease. It would be interesting to know if such an option effectively leads such companies to a better preventive policy.

4.1.5 Specific financial aid

The Finnish report emphasises that there is no State financial aid towards health and safety and that there is no reason to establish a system of aid in so far as this domain is entirely the responsibility of the company. The Danes also appear to think this way, though the philosophy is not shared by every country. In Germany, France and Sweden, there are specific funds whose purpose is clearly not to help in bringing things up to standard, but to support or encourage exemplary action which could then be imitated.

In **Germany**, the Arbeit und Technik (Labour and Technology) section attached to the Ministry of Labour can intervene financially to the total extent of between 30 and 42 million ECU. Aid is given to projects carefully selected for their significance and the quality of the lessons that could be drawn from them. Other aid is granted by the German Länder, the rapporteur estimating that this amounts overall to 5 million ECU.

In **Sweden**, the Working Life Fund which has been in existence since 1989 and, in principle, should cease operation in 1995, has a budget of 1.7 billion ECU which is provided out of a specific tax levied on business enterprises. The purpose of this fund is to assist with implementation of experiments encompassing all the factors which contribute to better working conditions. The fund looks after a share of the investment required for such experiments. The fund also provides an advisory service for small and medium-sized companies.

In **France**, apart from the activities of the health insurance funds and the Mutualité Sociale Agricole (agricultural mutual welfare scheme), a Fund for the Improvement of Working Conditions (FACT) has been in existence since 1978, whose objectives and mode of operation are similar to those of the Swedish Working Life Fund. The FACT undertakes to pay part of the cost of researching and implementing action over and above that stipulated by standards. This fund, whose overall budget is modest, at 431,895 ECU in 1992, is administered by the Ministry of Labour.

It would be interesting to have an evaluation of the results obtained through this form of aid.

Where aid exists in the other countries, the rapporteurs have been unable to identify its source or amount.

The aid we have just been discussing is actually intended for a small number of companies and represents only a small share of the total amount allocated by individual states for preventive action in the area of health and safety or, on a broader scale, the improvement of working conditions.

4.2 Knowledge and its diffusion

In the field which concerns us, the effectiveness of any action taken depends greatly on knowledge of the cause of occupational injuries and disease. Only the larger companies are likely to have the means required to develop specific research. In all the countries, the State plays a decisive role in collecting data, encouraging research, helping to distribute information and associating itself with training.

4.2.1 Data collection facilities

In 1992, the Foundation conducted research on this particular topic in twelve European countries, so there would seem to be little point in reproducing the details of information which is already available elsewhere. Accordingly, we will simply be examining a few noteworthy features revealed in the course of processing the data provided by the national reports.

Between the countries which participated in this research there are considerable differences in the facilities for collecting data on occupational injuries and, more especially, occupational diseases. Employers everywhere are obliged to declare occupational injuries under the terms and conditions stipulated by law. With the exception of the Finnish report, which asserts flawless recording of all injuries, all the reports claim that facilities could be improved. Where data exists in some countries, it is not yet centrally held and is not subjected to any specific form of processing. Sources of error due to omission persist even in the oldest systems, as indicated by the British and Danish rapporteurs. System quality also depends to a large degree on organisational structures and some highly distributed systems such as found in Italy can only produce relatively reliable data on the best organised regions. Occupational diseases are even more difficult to record, even in countries where the list is limited. The difficulties encountered are attributable, no doubt, to the possible absence of systematic data-collection procedures, but also to the quality of the information which heads of companies provide on injuries, or which medical officers submit on occupational diseases.

Despite the imperfections of data collection, the results provided are essential instruments when it comes to any form of policy-making. For instance, the observation in France of an increase in injuries over a period of two or three years very quickly led the Ministry of Labour to set up an **Occupational Hazards Monitor** (Observatoire des Risques Professionels) to find out the reasons for this deteriorating situation.

In the countries with the longest-established preventive policies, fast and systematic processing of the statistics provided both by ministries of labour and by insurance organisations has been accompanied, over about the past ten years, by systematic surveys conducted among samplings of employees. These surveys may take place every year, as in Sweden, or at regular intervals, as in France, Finland, the United Kingdom, the Netherlands and Germany. They can serve to validate the quality of data collected on injuries and diseases; for instance, it has been estimated from a survey of this kind that only a third of injuries are officially recorded in the United Kingdom. Above all, such surveys provide information on the employees' assessment of their working conditions and the nature of the hazards confronting them. Because this data is repeatedly produced, it allows one to get the measure of trends and developments in the way working conditions are experienced.

Questionnaire-based surveys are also used to acquire more precise knowledge of certain occupational hazards (Finland, Germany, Spain, Sweden, France, the Netherlands) or working conditions in specific industries or sectors. The survey method may be used in some companies and is compulsory in some cases: Finnish legislation even obliges the employer to monitor systematically and continuously the risks incurred by his employees. Finland has also devised a method of data collection enabling the measurement of stress. The Netherlands is conducting a survey on the same topic, the results of which should be known in 1995. Germany is currently giving 15.6 million ECU towards the creation of a **national health table**, as already used in some of the German Länder. This body of data is an essential tool for any form of health and safety policy and the facilities provided by computer media should make it accessible to a wider audience, as is already the case in Sweden and Finland, for example.

There would seem to be an urgent need to improve existing systems, which can only be done if the social partners and in particular the employers take note of the need to pass on their information. All European countries without such monitoring facilities must also be assisted to acquire them as rapidly as possible.

4.2.2 Research policies

The collection of data and surveys on working conditions are an integral part of any effective research facility.

The public authorities appear to play a decisive role in directing and coordinating health and safety research. Among the countries of the European Union, the differences in this domain are considerable. There is a variety of reasons for these differences, which are largely attributable not only to seniority of industrialisation and the means available for this type of activity (not always an obvious priority in the eyes of the man of action), but also to a general philosophy regarding the place of health and work in the development of the individual, which varies greatly from North to South in Europe, despite a very evident trend towards uniformity of behaviour.

● **Coordination**

Some countries, namely Sweden, Finland, the United Kingdom and Denmark, have established research coordinating facilities, either entrusting this task to their ministries of labour, or creating specific funds the scale of which effectively helps the coordination of research programmes.

In Sweden, the Work Environment Fund initiated by the social partners in 1972 has a budget of 55 million ECU per year to finance research projects conducted by the universities or the competent laboratories. The Fund is replaced by the Swedish Council for Work Life Research 1 July 1995. The Council will have at its disposal an annual budget of about 28 million Ecu (Skr 260 million) for research grants in allocation from the state budget. The Council will continue in the tradition of the Fund and work closely with the social partners.

In the United Kingdom, the Ministry of Research itself administers contracts agreed with 230 teams in an amount of £38 million per year. This is not, of course, the only ministry to finance research into health and safety, but its own internal Research Strategy Unit plays a dominant role in the area of coordination.

In Finland, the Government Council for Research and Technology performs this role of coordinating research between specialised institutes and the universities. The budget allocated to state research into health and safety borders on

25 million ECU, which represents about 0.03% of GNP and **11.6 ECU per employee per year.**

In Denmark, the task of coordination is incumbent on the Working Environment Fund, which has a budget of 2.3 million ECU.

As provided for in the legislation instituting France's National Agency for the Improvement of Working Conditions, this agency has been given specific terms of reference in the area of research coordination. These terms of reference have never been complied with because of lack of means and man-power. In Belgium, a working group is currently reflecting on the facilities which need to be put in place to ensure coordination of this kind.

The public authorities in the other countries have no means of directing and encouraging research for the very reason that they have no formal mechanism for coordination. In all countries, the ministry of labour provides a great deal of stimulus in this domain. In a number of countries, specialised institutes play a determining part in research as well as in the provision of information and training.

• Specialised institutes and the universities

All the countries with a long tradition of industry, except Belgium, have established specialised institutes for research into health and safety. Some of them go back a long way, like the INRS in France or the SfS in Germany, both set up just after World War II.

These institutes generally specialise in specific problems, such as health and safety, industrial medicine, or working conditions in general. Worth mention by way of example would be the new Swedish institute of work life and occupational health research (Arbetslivsinstitutet), merged from the Institute for Work Life Research (ALFI) and the National

Institute of Occupational Health (NIOH), the TNO in the Netherlands, the BAU and the BAfAM in Germany, the Higher Institute for Health and Safety at Work (ISPESL) and so on.

The universities naturally play an essential role in this system, especially the medical universities or schools (colleges of public health in Spain). Somewhat broadly speaking, no doubt, the institutes and the universities could be seen to share their roles inasmuch as the former conduct applied research, the latter basic research. These two types of research are defined very ambiguously and variously from one country to the next. In Sweden and Finland, applied research very often means a research campaign and an experiment or trial. Applied research can also be held to mean the development of tools and methods derived from the results obtained in basic research, as is the practice of the INRS in France.

Another distinction may be made between these two bodies under the criterion of the number of disciplines involved in their research teams. Even if there are exceptions, basic research tends more to be monodisciplinary because of the specialisation it demands. The institutes, on the contrary, bring together specialists and research workers from every discipline capable of dealing with the specific problems of health and safety in all their aspects.

With the ministries of labour, the institutes play a fundamental role in collecting data and making it available to researchers. Sweden and Finland, with databases on CD.ROM, have clearly developed a major lead and could serve as a point of reference in this area.

Our field of investigation should not exclude the research carried on within large companies, especially state-owned enterprises. As a general rule, and depending on the quality of the information available, this research is conducted on a contractual basis by specialised research agencies. Also to be taken into account is the sometimes very significant research carried out by the insurance services. Quite a few institutes are in fact directly financed by these bodies.

The trade unions can sometimes do their own research in the specific area of health and safety. Generally, these organisations involve themselves in the tripartite administration of some institutes, which also gives them a practical means of directing a part of the research activities.

This brief general view of the policies employed in the area of research still leaves several questions unanswered. The association between the results of research work and decisions actually taken is always a problem. Specific research into toxic materials, for example, has enabled considerable progress to take place in protecting against and often eliminating such products. But when it comes to human behaviour, the recommending of measures is a far more delicate matter.

The most difficult question concerns the very definition of research topics and their application to specific problems encountered at company level. The participation of the inspectors of labour and the medical officers, and also the social partners, is indispensable in elaborating questions which may appropriately be addressed to research. Several reports underline here the gap which sometimes exists between certain research work and problems in practice. No doubt there is a need to maintain what is always a difficult balance between basic research and applied research.

Several reports also mention the reduction in funding of research (note the closure of research centres in the United Kingdom). The temptation is naturally very strong, in times of economic crisis, to cut back on money allocated to research; as the trade unionists say, a research workers' strike never really works! The really important problem, scarcely touched on by any of the reports, is that of training the research workers and safeguarding a research potential which takes a long time to establish.

Finally, the decisive role of European aid in developing research in countries in which it still receives too little attention should be noted.

4.2.3 The distribution of information and sensitisation

As with research, to which it is often closely linked, the distribution of relevant information assumes very different proportions from one country to the next.

• Information provided by employers

With the possible exception of Ireland, Greece and Portugal, all the countries have exceptionally precise and detailed legislation regarding the body of information to be provided by the employer to the ministry of labour, the employees' representatives and the employees themselves. The list of these obligations appears to be the outcome of a succession of regulations adopted over time in response to problems requiring resolution. Their sum total is nonetheless impressive and should allow very detailed knowledge of what goes on in these companies. Likewise, the information which employees provide pursuant to the legislation in most countries should be remarkable. The question one would like to ask concerns the use that is made, in most cases, of the information which reaches the social partners and the inspectorate of labour. Its use depends, of course, on the means at the disposal of the public authorities to process it systematically, as well as on the competence and the forcefulness of the employee representatives in making appropriate use of it.

The provision of information to employees on their working conditions at the time of recruitment or when changing jobs is also a legal obligation in almost every country. How is observance of this obligation controlled? Whose responsibility is it to give this information and how is it communicated? How capable are employees of receiving it? So many questions, to which the reports give only a formal and far from satisfactory answer.

- *Information provided by ministries, institutes and professional associations*

If added together, the list of reviews, reports, information leaflets, audio-visual media, etc. published in the various countries would amount to several score pages.

Documents on health and safety or prevention at the workplace are generally well-targeted. They deal with varyingly specific topics and are aimed at clearly identified audiences, as follows:

- Information from the ministry of labour on the laws in force.
- Information from employee representatives on the laws in force, but also on policies and practices to be developed. Documentation of this kind can come from specialised institutes, although also from trade unions who are generally highly sensitised to the problems of health and safety.
- Information intended for company medical officers and those in charge of safety at company level. This documentation often presents the results of research work.
- Information for scientists, essentially concerning the results of research.

The insurers may also involve themselves in this production of information. In Germany, for example, the Berufsgenossenschaften (mutual indemnity associations) provide branches of trade and industry with abundant sectorial information on problems specific to them.

This information, part of which is free, is paying off in an ever-increasing number of cases at the present time. While produced in excess in some countries, it still seems very insufficient or indeed non-existent in others.

The information is not limited to written material. The European Year of Health and Safety (1993) gave every country the opportunity to mobilise intensively on this topic; meetings, debates, symposia and presentations of experiments punctuated the year and may well have made a greater contribution to creating awareness of these problems than the whole body of publications otherwise distributed. Apart from that favoured occasion, various countries conduct national campaigns on television. The 'health circles' established in Germany are especially deserving of attention.

The ministries, institutes, insurance organisations, etc. do not just content themselves with issuing information, they also respond when its provision is requested; by way of example, the Health and Safety Executive in the United Kingdom replied to 750,000 requests for information in a period of a year.

4.2.4   Training

In many ways, training is part of the continuity of information. Information given to an employee on his or her workplace is ineffective if it fails to provide real training. Employee representatives cannot avail of the abundant information they receive unless they have sufficient training to do so.

• The training of experts

The majority of countries operate a policy of training experts (meant here by experts are the inspectors of labour and medical officers, safety officers or the employee representatives). Some countries provide a very significant volume of training of this kind, both initial and on-going. Community aid would undoubtedly be needed to enable its development in countries where health and safety are not yet a priority. Taken overall, it is certainly not on the training of experts that attention would need to be focused, even though there is still room and a desire for improvement.

• The training of employees

Sensitisation to the questions which concern us in this paper is reliably developed through early education, though this is non-existent almost everywhere. Finland is in course of discussing a project for secondary schools. The teaching of health and safety recently became obligatory in Sweden, where the schools have the necessary training materials, though the use made of them is not yet known. In the best of cases in the other countries, training is provided in the secondary schools for vocational education. Training of this kind is also available in the engineering colleges where, quite often, it only occupies a very small part of the syllabus. But what of the other forms of training provided by the universities, also aimed at training the future heads of enterprise? At that level, training can only develop if research is also carried on and is capable of producing teachers-cum-researchers.

Employees necessarily receive their information and training through their foremen and management. None of the reports makes mention of training specifically designed for that particular group, which nonetheless occupies a key role in the matter of prevention.

The point which emerges most clearly from a reading of the national reports is that knowledge of the problems concerned, research, information and training are closely associated. These elements constitute a system and cannot be dealt with in isolation while ignoring the interaction between them.

It is under this heading, too, that one can see the most important differences between the countries of the European Union, and the slow development which the latter could help to overcome.

While information and training should, of course, retain a national flavour, the results of research deserve to be shared more systematically than they are today, placing them at the disposal of countries which have not had the means to develop their own independent research on the topics which concern us. Comparative research should also assume a more important position in future.

CONCLUSIONS

Several conclusions emerge from this consolidated report. The first is an overall view of the situation as it stands. The second concerns an assessment of system performance and the last relates to the vision of the future expressed by those consulted by the rapporteurs.

• **A contrasting situation**

Of necessity, the gap is considerable between the countries whose concerns in the area of health and safety are expressed in provisions some of which are now more than a hundred years old and those who are beginning to install the infrastructure they need to address these questions. These contrasting situations explain to a large degree the differences in levels of concern which show through the reports. In the countries which are less advanced in this area, health and safety is an objective in itself. For others, and this is especially true of the Scandinavian countries, Germany, the Netherlands and perhaps to a lesser degree, the United Kingdom and France, **the concepts of general health and wellbeing at work enrich and enhance the way in which the problems of health and safety are perceived and open up new prospects for action.** No longer is it simply enough to concentrate on preventive measures aimed at reducing occupational injury and disease, now there is also a need to design ways of working which will avoid 'wear and tear' and allow genuine advancement in health. As emphasised in the Dutch report, the spectre of a society of 6.5 million workers with a million invalids as its responsibility compels reconsideration of the way in which health and safety is approached at company level.

This contrasting situation also explains reactions to European Directives. In the countries in which health and safety are a recent concern, these Directives are very welcome to the extent that they provide a ready-made framework and help to accelerate the installation of the infrastructures required to put them into practice. In other countries, while they may meet with a generally favourable reception, they give rise to various kinds of anxiety.

The enacting of regulations by the competent authorities of the EU could well weaken national bodies who have been traditionally entrusted with the enacting of rules. This anxiety is manifest in the German report in relation to the prescriptive role of the insurance schemes.

This situation is leading to the development of new, very active structures within the CEN (European Standards Commission) and the CENELEC, as referred to in the French and German report.

Concerns over a deterioration in the system are significant in countries in which the level of standards already achieved is greater than that stipulated by European norms. The maintenance of higher standards certainly involves the risk of competitive disadvantage for the companies adhering to them and would therefore lead, in the long run, to a worsening of the situation. Finally, some countries, the United Kingdom in particular, fear an increase in the body of regulations at the very time they are pursuing a policy of actually simplifying or even eliminating rules, although such measures have hardly touched on rulings in the area of health and safety.

In contrast, all the reports mention the benefits of European regulations which also apply to the public sector and which insist on participation and training.

• A difficult assessment of system performance

The effectiveness of a system can be measured by the results obtained. The quantitative data the reports provide on occupational injury and disease is difficult to interpret and, as we had occasion to emphasise in our introduction, cannot always be compared.

Taken overall, the most recent figures indicate a significant reduction in industrial injuries (except in Italy). Very often, this reduction follows an increase concomitant with the resumption of activity and is contemporaneous with a growth in unemployment. Accordingly, it would seem to be the sole virtue of unemployment that it results in a reduction in accidents at work! Apart from the effectiveness of preventive measures, these results are contributed to by changes in manufacturing industry and the relative importance of the manufacturing and services sectors. Occupational diseases may be changing in nature but, in general, they remain virtually stable in number. **In fact, it is quite clear that the available data collecting and processing systems are unable to provide correct information on the reality of the situation.**

Table 3, which presents a summary of selected data on industrial accidents, should be interpreted with great caution. With the exception of Finland, no country pretends to exhaustive coverage of all injuries. Errors of communication can also intervene. The regulations governing declaration vary from one country to another according to seriousness which can be measured by the number of days' stoppage as a result of an injury. Some countries, like Sweden, have just changed their rules of registration. Not only is it impossible to make comparisons, the figures provided represent reality to a varying degree, greatly underestimating it in some cases (the Irish report estimates that 95% of occupational diseases are not recorded). Finally, the raw data provided by some rapporteurs incorporates all injuries in the total number, irrespective of cause, while other rapporteurs have rightly distinguished injuries at work from those in transit.

Table 3: Number of injuries (latest figures presented)

Country	Fatal Injuries	Total Injuries
Germany (92)	1310	2,069,422 (55/1000)
Belgium	not provided	
Denmark (93)	61	44,247
Spain (90)	1446	700,000
Finland (91)	not provided	112,500 (of which 13.8% in transit)
France (91)	1821 (of which 40.6% in transit)	951,517 of which 10% in transit)
United Kingdom (92)	249	168,383
Greece (92)	120	25,063
Italy (91)	1858	1,020,000
Ireland (93)	64	3606
The Netherlands	56 (91)	170,000 (28/1000) (93)
Portugal (93)	185	278,455
Sweden (93) (94)	108 214[*]	41,319 39,190

[*] of which 129 in the Estonia catastrophe

These figures in themselves provide a legitimate reason to continue the efforts already made.

The hazardous sectors are well known and are the same in all the countries. The construction industry and metal working come top of the league everywhere but agriculture, since its mechanization, has also paid very high tribute to injury, especially fatal injury (1/3 of the latter in Ireland). These injuries occurring in industries in which the rate of female employment is low chiefly affect males. Some reports provide more precise data on age, status and level of training (e.g. France). Although one cannot generalise, those proportionately most affected are young workers with the least training, temporary workers and foreigners; this is also because they work in the most hazardous sectors of industry.

Small companies experience a higher accident rate than large ones.

The data on occupational diseases reported is very difficult to put in perspective, because the reports are very mixed in their forms of presentation. Here again, the major constants appear in connection with the principal diseases whose classification differs little from one country to another; periarticular ailments, dermatoses and deafness mostly come at the top of the list and the material most criticised is asbestos. The German rapporteur is the most critical when he considers, no doubt with some reason, that **the occupational diseases identified today reflect the working conditions of the past. The lists of recognised diseases are probably obsolete** and this obsolescence is a serious handicap in the implementation of effective preventive policies. The same rapporteur likewise disputes the simple causal relation made between exposure to a risk and identification of disease whereas, in fact, changes at work result in a variety of exposures to nuisances whose effects are very little known. It has to be said, on that point, that what could be described as national inflexibilities prevented the implementation of the European recommendations of 1962 and 1968.

Employees, when asked, indicated awareness of a deterioration in their working conditions, though it is still very difficult to interpret the findings of these surveys.

Irrespective of the considerable cost to business enterprise and to society which, in any case, should never be measured against human hardship, a determined effort should be made to improve the collection and harmonisation of data which is the raw material on which policies can not only be built, but also evaluated and reoriented.

● **A vision of the future, strengths and weaknesses**

Of the experts consulted by the rapporteurs, none presents an overall vision or a model made up of the facilities and means which ought to be put in place within the next ten years with a view to developing effective policies in the area of health and safety. The scope of the survey was too restricted, no doubt, to allow the experts consulted to develop all their ideas. On the other hand, all of them noted what seemed to them the strengths and weaknesses of the systems in place and the elements they thought deserving of particular attention.

- The strengths

● In almost all the countries, there is a very strong consensus of opinion between the public authorities and the social partners on the importance to be attached to the subject of health and safety. Despite a difficult economic situation, no voices were raised in favour of any reduction in preventive and protective measures. Everyone emphasised that this subject constitutes a major point of cooperation between management and unions.

● Even if there are some complaints about the overall operation of the inspectorate of labour, the sterling quality of the individual inspector is recognised in all countries in which the service has a long-established tradition. The Danish report even makes him the key-man in improving working conditions and safety in that country. Also underlined is the importance of the inspectors' independence vis-à-vis all other parties.

● As a general rule, the very favourable assessment of the quality of the inspectors of labour is likewise accorded to the experts who are consulted or intervene under various headings in connection with various problems.

- The weaknesses

• The first weakness is in the available means in manpower, facilities and infrastructure, a patently obvious fact in Ireland, Greece and Portugal although also arising elsewhere, especially with regard to manpower. In a context not only of economic crisis, but also of deregulation, budgetary restrictions are severely felt and can even jeopardise the inspection services, which are not overabundant in any country.

It is not the personnel or quality of personnel, but the underlying structures and models which for various reasons are in crisis.

• The second weakness, largely escaped by the Scandinavian countries and Belgium, is the weakening or weakness of the social partners. The reasons for this are many and it is not the place of this report to discuss them. However, several national reports emphasise that this weakness poses a very definite problem for health and safety representatives, who turn to the unions for the information, training and advice they need to perform their function.

• There is agreement on the importance of training, which ought to begin at school, even if teaching programmes on health and safety are still the exception. If there is still room for improvement in the training of experts, the real need for development is in the appropriate training of management and employees, still seen throughout as highly inadequate.

• SMEs are the blackest point observed in every country. The structures, actors, legislation and controls in place have in point of fact been conceived for larger companies, so consideration of measures applicable to SMEs may be seen as a priority.

• Finally, some experts mention the lack of innovation in research and regret the research workers' insufficient concern for applied research.

• **Lines of future thought and action**

These are many and several have already been mentioned. The German report, which undoubtedly presents the best-developed conclusions, questions **the future of preventive measures which have proven their worth in a given context but may, at the present time, be at the end of their effectiveness.** The health circles being established in Germany and the participative methods on which they are based could be an important way forward.

The traditional risk-factors have been well identified, but it remains to explore and take note of the diseases occurring in the services sector and post-industrial society. While the intensification of work, stress, the discrepancy between levels of training and the content of jobs on offer, the growing proportion of women in the working population and the ageing of the population are equally deserving of research in themselves, they also call for the establishment of a different approach to action. The organisation and content of work can no longer be ignored in policies aimed at prevention. Which also means that a new body of experts should appear on the health and safety scene and above all, that multidisciplinary research and concerted action should become the rule and no longer the exception.

As is also underlined in the German report, industrial policies should be developed and the **insurance schemes, no matter what their status, should initiate policies of prevention.**

The economic context has its effects on the means placed at the disposal of existing structures. It also has its more indirect and less easily analyzed effects on employees and their representatives, for whom employment is becoming a priority at the expense of working conditions. Independently of this economic context, though perhaps accelerated by it under the heading of flexibility, there is a development throughout of practices which are changing the relationship between employer and employee. Existing preventive and protective structures seem inadequate in the face of these developments. Some experts are also worried about the new ways of arranging working hours which pose specific problems not yet faced in the area of safety and health.

Apart from the criticisms or reservations reported above in the area of health and safety, Europe appears as an opportunity, allowing countries less advanced in this field to catch up very rapidly and also serving as the essential forum for a sharing of knowledge and experience. It is no less true to say that the approaches observed reflect different perceptions of the role of work in our lives, traditions which should be respected if we are to bring about a real improvement in health and safety conditions.

European Foundation for the Improvement of Living and Working Conditions

Policies on Health and Safety in Thirteen Countries of the European Union
Volume II: The European Situation

Luxembourg: Office for Official Publications of the European Communities

1996 – 102 pp. – 21 x 29.7 cm

ISBN 92-827-6642-X

Price (excluding VAT) in Luxembourg: ECU 11.50